CARRIAGE BARNS

Sources of Building Plans, Kits,
Products and Services to Help You Create
a New Garage, Workshop, Stable, Backyard
Office, Studio or Live-In
with Old-Style Charm

Edited By Donald J. Berg, AIA

ISBN 0-9663075-3-4
Library of Congress Catalog Card Number: 99-076562
Copyright 2000 by Donald J. Berg

Published by:
Donald J. Berg, AIA
PO Box 698, Rockville Centre, NY 11571
516 766-5585

E-mail: djberg@aol.com

The use of quotes and brief passages by reviewers or in articles crediting this publication is welcome. Please call the publisher for permission for any other use. Many of the illustrations and plans used in this book are the copyrights of the individual designers and plan services credited. Please contact them directly for permission to use their illustrations.

Efforts have been made to ensure that the information in this book is accurate and that the products and services mentioned are of the highest standard. However, all of the designers, manufacturers and experts featured are individually responsible for maintaining the quality of their production. Recommendations and information are provided by the editor and publisher with no guarantee of accuracy or of suitability for your use.

Contents

Introduction...5

How to Use This Book...5

Restoring a Carriage Barn...7
Building a New Barn...7
Building Resources...8

Yesterday's Carriage Barns...9

Today's Carriage Barns...23

Stables...39

Live-Ins...45

Loft Garages...57

Cupolas...66

The Barn Building & Restoration Directory...67

References & Resources...89

Internet Resources...89
Building Codes...91
Building Book Catalogs...91
Periodicals...92
Barn & Backyard Building Guidebooks...92
Books on Yesterday's Designs...94

Index...96

Carriage House Illustration from *American Country Building Design*

About the Editor
Don Berg is a member of the American Institute of Architects and of the Society of Architectural Historians. His designs and articles have been published in *Traditional Building Magazine, The Old-House Journal, Home, Hudson Valley Magazine, Yankee Home* magazine, and many other publications. He appeared in the TV special, *The American Farmhouse* on the Home and Garden Network. Don has written or edited fifteen books on traditional building and landscaping.

Other Country Building Books by Donald J. Berg, AIA

AMERICAN COUNTRY BUILDING DESIGN: Rediscovered Plans for 19th Century Farmhouses, Cottages, Landscapes, Barns, Carriage Houses and Outbuildings

THE BACKROAD HOME: Simple Country Designs of Cottages, Cabins, Barns, Stables, Garages and Garden Sheds with Sources for Blueprints, Kits, Building Accessories, Catalogs and Guide Books

BARNS AND BACKBUILDINGS: Designs for Barns, Carriage Houses, Stables, Garages and Sheds with Sources for Building Plans, Books, Timber Frames, Kits, Hardware, Cupolas and Weather Vanes

COUNTRY PATTERNS: A Sampler of American Country Home and Landscape Designs from Original 19th Century Sources

THE DOOR YARD: Old Time, Common-Sense Advice on How to Plan, Plant & Maintain a Beautiful Home Landscape and
HOW TO BUILD IN THE COUNTRY: Good Advice from the Past on How to Choose a Site, Plan, Design, Build, Decorate and Landscape Your Country Home

Introduction

Since I published the book *Barns and Backbuildings*, I've gotten more than a hundred calls and letters from readers asking for carriage barn plans. I assumed, at first, that they were planning to build reproductions of classic 19th century carriage houses. I was surprised because where I live people rarely travel by surrey. Little by little, I learned that different people were hoping to build at least five entirely different types of backyard buildings that they all called carriage barns. I decided to add to the confusion with this book.

Most of the requests came from people who owned traditional houses. They were looking for garages that harmonized with their homes. Some were looking for the same traditional style in new horse barns. Others were planning to build barns or garages with living space upstairs. In certain parts of the country, the terms "carriage house" and "carriage barn" are synonymous with any building that combines a garage and an apartment. Some people were looking for authentic historic carriage house designs to convert for use as garages for the 19th century houses that they were restoring. One caller actually wanted a carriage barn. She had horses and a small collection of antique wagons.

I'm very lucky to have a group of friends who are talented designers and who offer their building plans to all by mail on the Internet. Most have helped me out before with other books. For this one, I simply asked each for their best carriage barns. The designs that they shared are the exact same mix of building types that people were asking for: traditionally styled garages, live-ins, stables, reproduction designs and one design, The Mokawk Barn, that, in plan at least, is a modern counterpoint of yesterday's carriage barns. I added a few of my own interpretations and a selection of authentic 19th century carriage barn designs, just for inspiration.

If you're planning to build a garage, backyard barn, stable, home office, studio or live-in, you'll find a variety of designs and information on finding even more in all of the designers' catalogs

Introduction 5

or on their websites. You'll also find information on inexpensive blueprints and easy-to-build kits. When you build, you'll find old-style hardware, cupolas, weather vanes, carriage house doors, woodwork, coach house lamps and many other hard-to-find products in this book's Directory. If you're one of the lucky few who have an antique carriage house or barn, you can use the Directory to find specialists who can help you restore and maintain it. I hope that there's something of use for you.

Thank You

I appreciate all of the plans, illustrationss and ideas that the architects, designers, and manufacturers shared with me for this book. Dave Noffsinger improved the whole book by preparing many of the illustrations of the new designs. My wife Christine helped with advice and editing. I'd like to thank her, my daughter Bethany and my sons Christopher and Ted for inspiring me by their remarkable successes at what they do.

In particular, I'd like to thank my dad, Winfred Berg. An engineer, designer and inventor, he tinkered and tweaked everything around our home to make things work better. He added a loft and carriage barn hatch door to our attached garage and that changed it completely. My brothers and I used the loft as a club house and hideout. The block-and-tackle ride up was always fun. Then, when Dad made the mistake of cutting a door from the loft to our bedroom, the ride down and out at night was always an adventure.

Don Berg, October 29, 1999

How To Use This Book

A Carriage-House and Stable design by architect Issac H. Hobbs, from his 1873 book, *Hobbs's Architecture*.

Restoring a historic carriage barn:

If you're lucky enough to be restoring an old carriage house, you'll find inspiration and hints about appropriate details by studying yesterday's designs. You'll find samples from 19th century architects' plan books, farm journals and mail-order plan catalogs in the next chapter. The Directory, starting on page 67, lists specialists who can help you restore and maintain your antique building and also has sources of reproduction or salvaged products for authentic reconstruction.

Building a new backyard barn, garage, stable, studio, home office, guest house or live-in:

Starting on page 23, you'll find designs of traditionally-styled backyard buildings. Most are flexible enough to serve a variety of uses. Construction blueprints are available for most of the designs and building kits are available for the others. If you find a design that you like, just contact the architect, designer or manufacturer directly for complete information on ordering construction plans or a complete kit.

The floor plans in this book are all drawn at the same scale: 1/16" = 1'-0". The smallest notch on a standard ruler is a foot on a floor plan. You can quickly compare the sizes of the different designs by comparing the plans. When you find one that you like, you can check how your cars, trucks, boats, equipment or furnishings will fit by measuring them and comparing the measurements to the spaces on the plan.

Mail-order blueprints are prepared to be used at sites with average soil and weather conditions. They may need to be adapted for use on your property and to meet the requirements of your local building and zoning ordinances. All mail-order plans, and the contracts and specifications of kit manufacturers, should be reviewed by a building professional who is knowledgeable about conditions and ordinances in your area. Contact your community's building official for information on required permits.

The Directory lists additional designers, builders and manufacturers. Review their listings for additional sources of designs. You'll also see listings for manufacturers and artisans who can provide you with timber frames, hardware, building materials, cupolas, weathervanes, coach house lanterns, barn doors, rolling door tracks and much more. If you're planning a horse barn, you'll find listings for stall partitions, flooring, ventilators and other equipment.

Building Resources

Starting on page 89, you'll see information on books, references and on-line services that should help you as you plan your project.

A Design for a Stable, by architect Calvert Vaux, from his 1857 book, *Villas and Cottages*.

Yesterday's Carriage Barns

The residence of Benjamin Moses, Seneca Falls, NY, 1876

Carriage barns were an only-in-America mix of function, fashion and folly unlike anything before or since.

In colonial America, carriages and carriage barns were not at all common. They existed in cities and among the elite. "Carriage trade" is still a synonym for the wealthy. Farm work wagons served also as the family transportation for country folk. City dwellers depended on hired livery wagons. But, in the 19th century, industrialization and the progressive farming revolution brought a combination of prosperity and mass manufacturing that made carriages affordable to many. By the 1850s, light, fast and elegant buggies were joining the old wagons on the farms and becoming essential family vehicles in villages and in the growing suburbs.

Unlike the sturdy wooden farm wagons, carriages needed shelter from the weather. Their bodies were usually just painted leather or fabric stretched over a light frame of wood and metal. At first, people just extended their barns or homes with tacked-on, open carriage sheds, but soon a new type of building evolved.

The most common type of carriage barns built in America were small wooden buildings that would house horses and wagons and might also shelter chickens, a cow and yard tools. Plans were almost always a simple rectangle, divided in half with one side for animals, feed and tack and the other as a "carriage room." They were called "wagon barns," "stables," and "horse barns" but the carriage room defined their purpose. These new structures were essential utility buildings, but since they were the largest backyard buildings on suburban and village properties and the ones that, for convenience, were usually built closest to the house on farms, they were very important to the look of a homestead.

People tried to make their carriage barns fashionable. They decorated them with bits and pieces of the latest building styles. Classic columns, pointy gothic gables, gingerbread trim, roof brackets and cupolas were common on 19th century homes. Different details were essential to each of the classic revival styles of the Victorian age. Their order, proportion and symbolism were serious subjects among architects of the day. However, when those same elaborate, high-style details were tacked on to a barn, it was hard to take them as seriously. Designers just seemed to have fun with carriage barns.

When carriage barns were popular, most homes were built from published plans and mail-order blueprints. Nineteenth century architects' pattern books and mail-order catalogs show thousands of home designs but very few carriage barns. In fact, the samples shown on the pages of this book are probably more than a third of all of the designs ever published. So, most carriage barns were designed by home owners and builders, and that probably explains their exuberance. Lewis Allen, a farmer, claimed to be a better designer of such barns because he was "free from the dogmas" of architectural schools. Many old carriage barns show the same spontaneity of design as the best Folk Art. They have wonderfully oversized cupolas, windows that are far too elegant for a barn and a jumble of different styles in the decorations.

The same industrialization that helped create carriage houses doomed them to a very brief life span. Less than fifty years after carriages became popular, the automobile rapidly replaced horses as America's transportation of choice. The garage, just as rapidly, replaced the carriage barn. Unfortunately, garages were less fun. Without lofts, scuttles, ventilators, hay hatches and all of the windows and doors needed for both horses and vehicles, they could be seriously simple. The first ones were described by Holly Wahlberg, in *Old House Journal*, as "pupose-built...12'x18' rectangular boxes." Today most garages are bigger, but still just efficient and somewhat dour rectangular boxes.

Page through some designs of yesterday's carriage barns for a glimpse at what we've lost.

A Carriage Barn design by architect R.H. Robertson, from the magazine *American Architect & Building News*, November, 1888.

A Wooden Stable for Three Horses, by George Harney, from the book Barns, Outbuilding and Fences, 1870.

Cold Spring, New York architect George Harney designed elegant custom homes, throughout the Hudson Valley, for a wealthy clientele. Many architects of his day published books of their best home designs to promote their services. Harney decided to do something different. His 1870 book, *Harney's Barns, Outbuildings and Fences*, featured just the backbuildings that he had designed for the homes. As the only architect's plan book that presented a large selection of carriage barns, it was very influential. You'll see many details inspired by Harney on old carriage barns across the country.

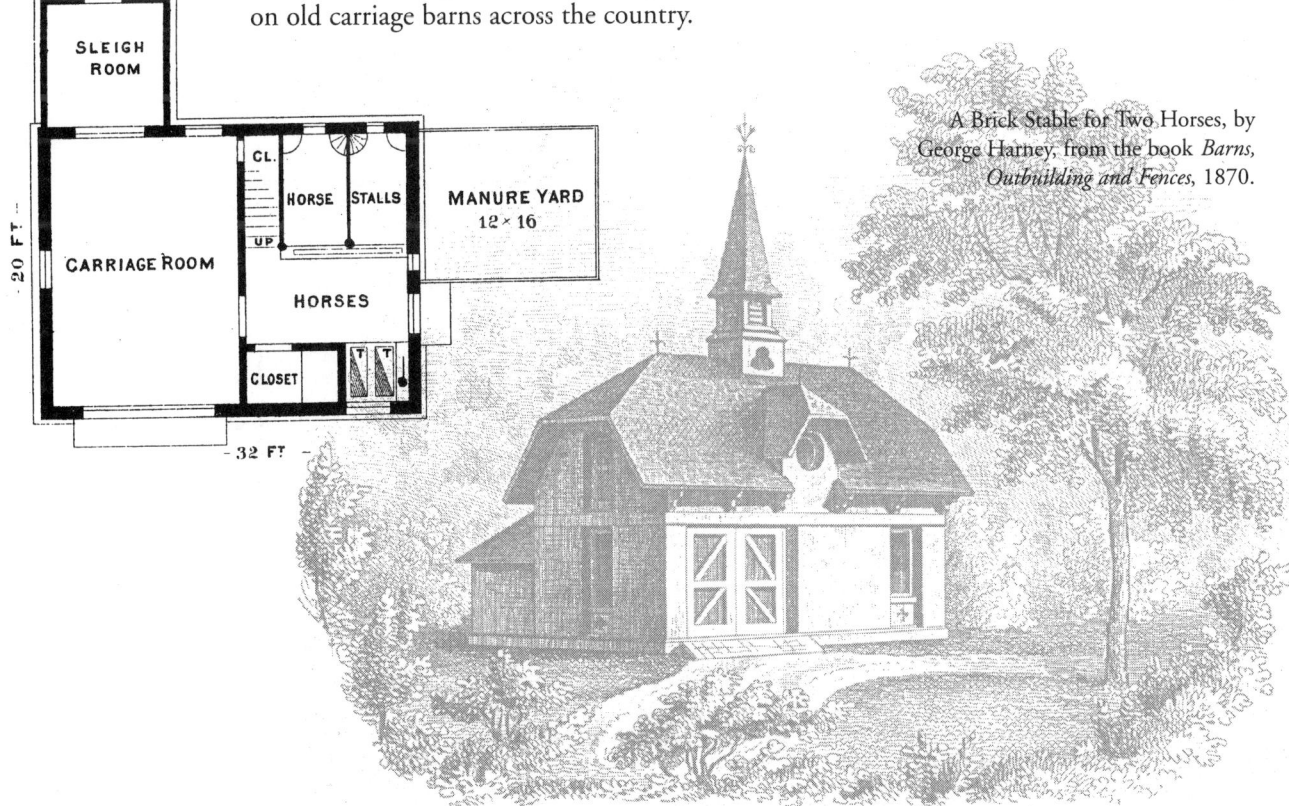

A Brick Stable for Two Horses, by George Harney, from the book Barns, Outbuilding and Fences, 1870.

Yesterday's Carriage Barns 11

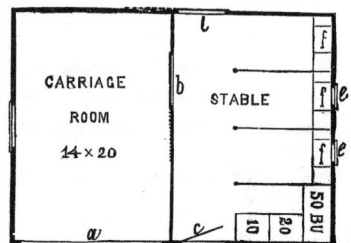

A Plan for a Small Barn, by Professor G.T. Fairchild of the Michigan Agricultural College, from *The American Agriculturist*, September, 1880.

Key to the notes on the plan:
a - 9' wide sliding door; b - Sliding door;
c - Stable door; d - Back door; e - Windows;
f - Feed bins, below open shoots from the loft.
Feed storage bins, shown with their capacity in bushels, are below the stairs to the loft.

Nineteenth century farm journals promoted carriage barns as an alternative to fragile tacked-on sheds and to the common, time-wasting practice of rolling carriages into the work area of barns at night and out again in the morning. Their drawings tended to show austere buildings, but farm families often built fancier versions from the same floor plans.

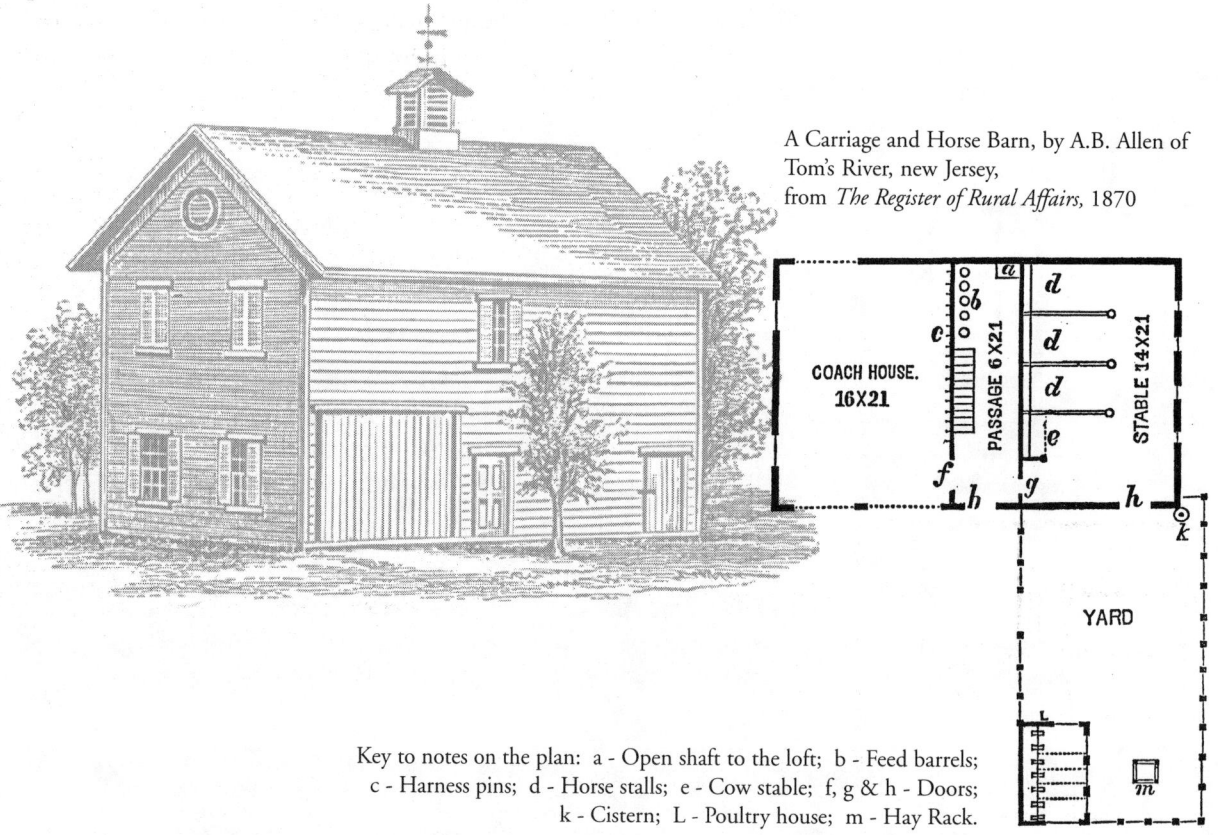

A Carriage and Horse Barn, by A.B. Allen of Tom's River, new Jersey, from *The Register of Rural Affairs,* 1870

Key to notes on the plan: a - Open shaft to the loft; b - Feed barrels;
c - Harness pins; d - Horse stalls; e - Cow stable; f, g & h - Doors;
k - Cistern; L - Poultry house; m - Hay Rack.

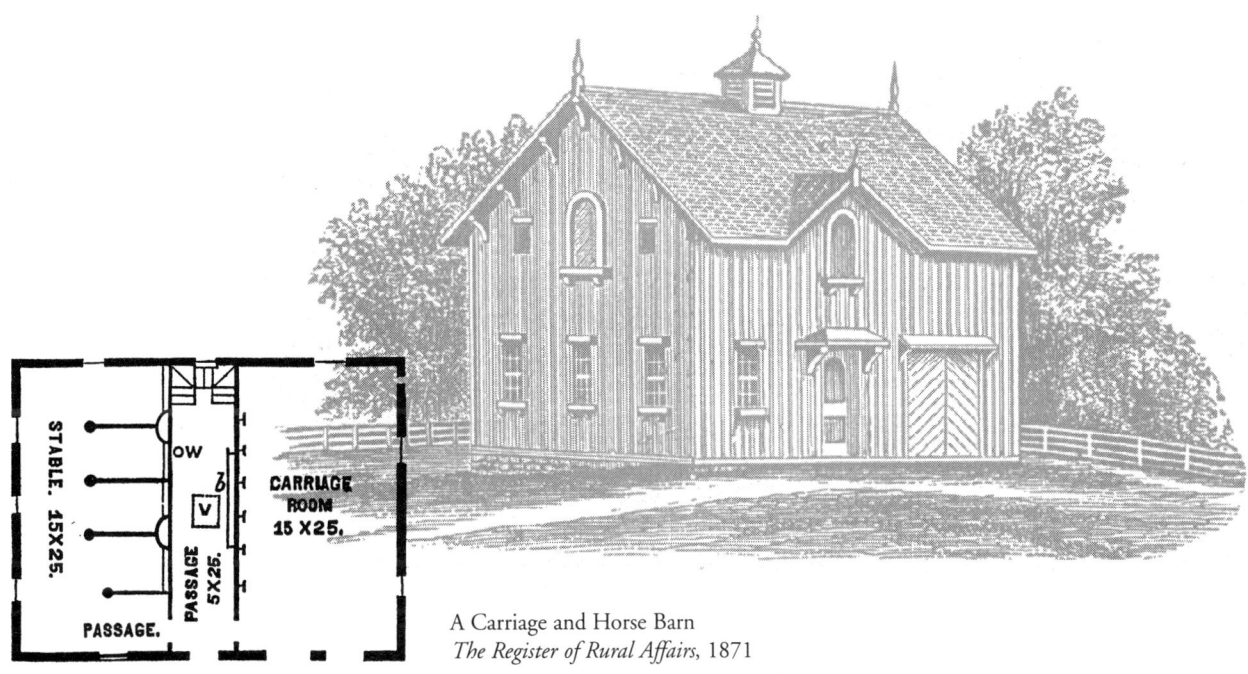

A Carriage and Horse Barn
The Register of Rural Affairs, 1871

The Annual Register of Rural Affairs was the *Reader's Digest* of its day for farm families. It promoted scientific farming and agricultural education by condensing some of the year's best articles from a number of popular farm journals. It also presented designs of innovative buildings, vehicles and tools. Editor J.J. Thomas was an exceptionally talented designer who had a way of creating elegant farmhouses, barns and backbuildings that served the practical needs of farmers. These two pretty carriage barns are typical of his style.

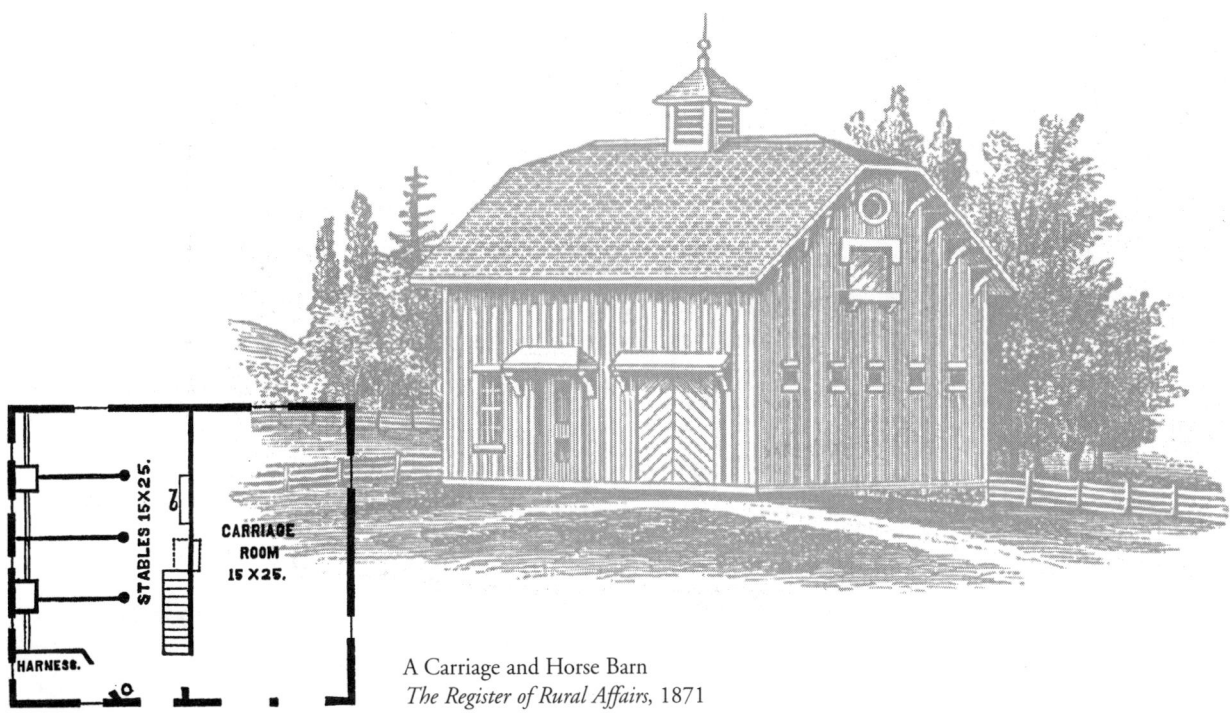

A Carriage and Horse Barn
The Register of Rural Affairs, 1871

Yesterday's Carriage Barns   13

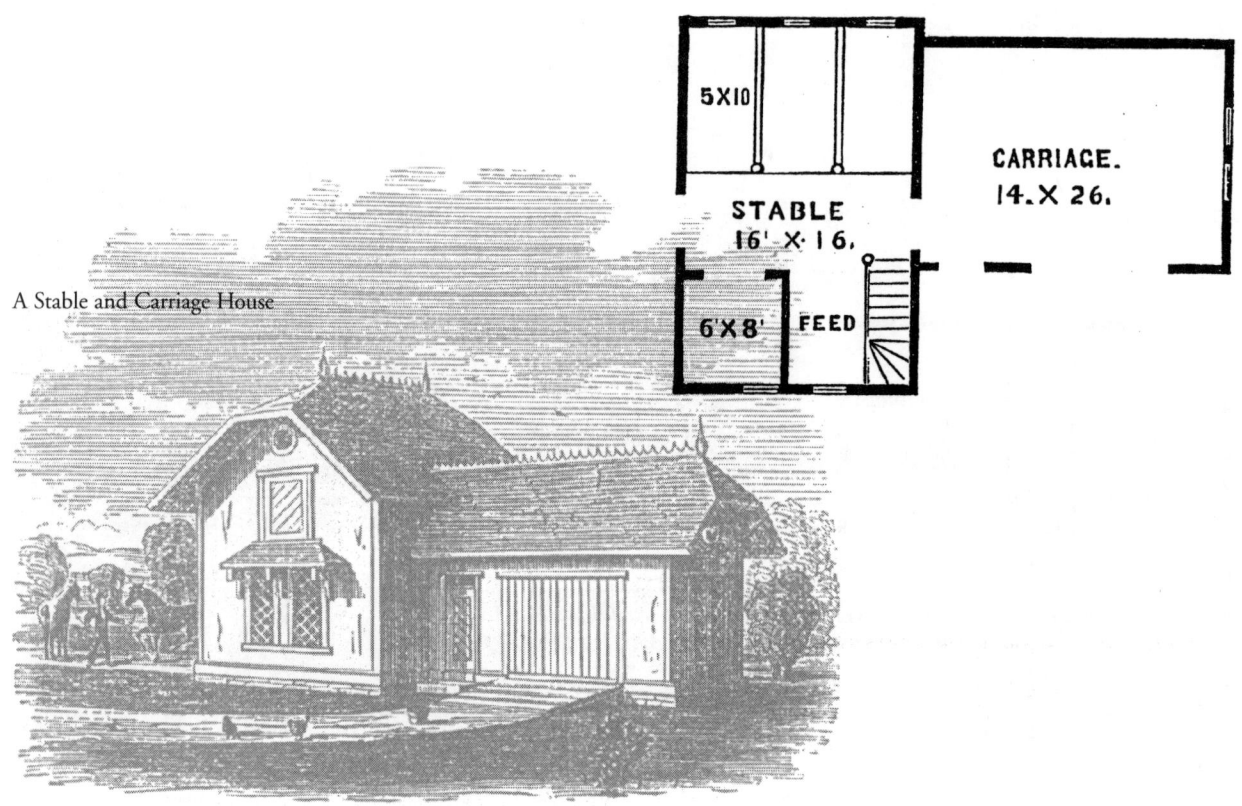

A Stable and Carriage House

Architects George and F.W. Woodward aimed their practice at clients in the new and growing suburbs. They published a series of plan books that presented designs of homes and outbuildings that blended the high-style of city buildings with the simplicity of farmstead designs.

A Cottage Stable

14 Yesterday's Carriage Barns

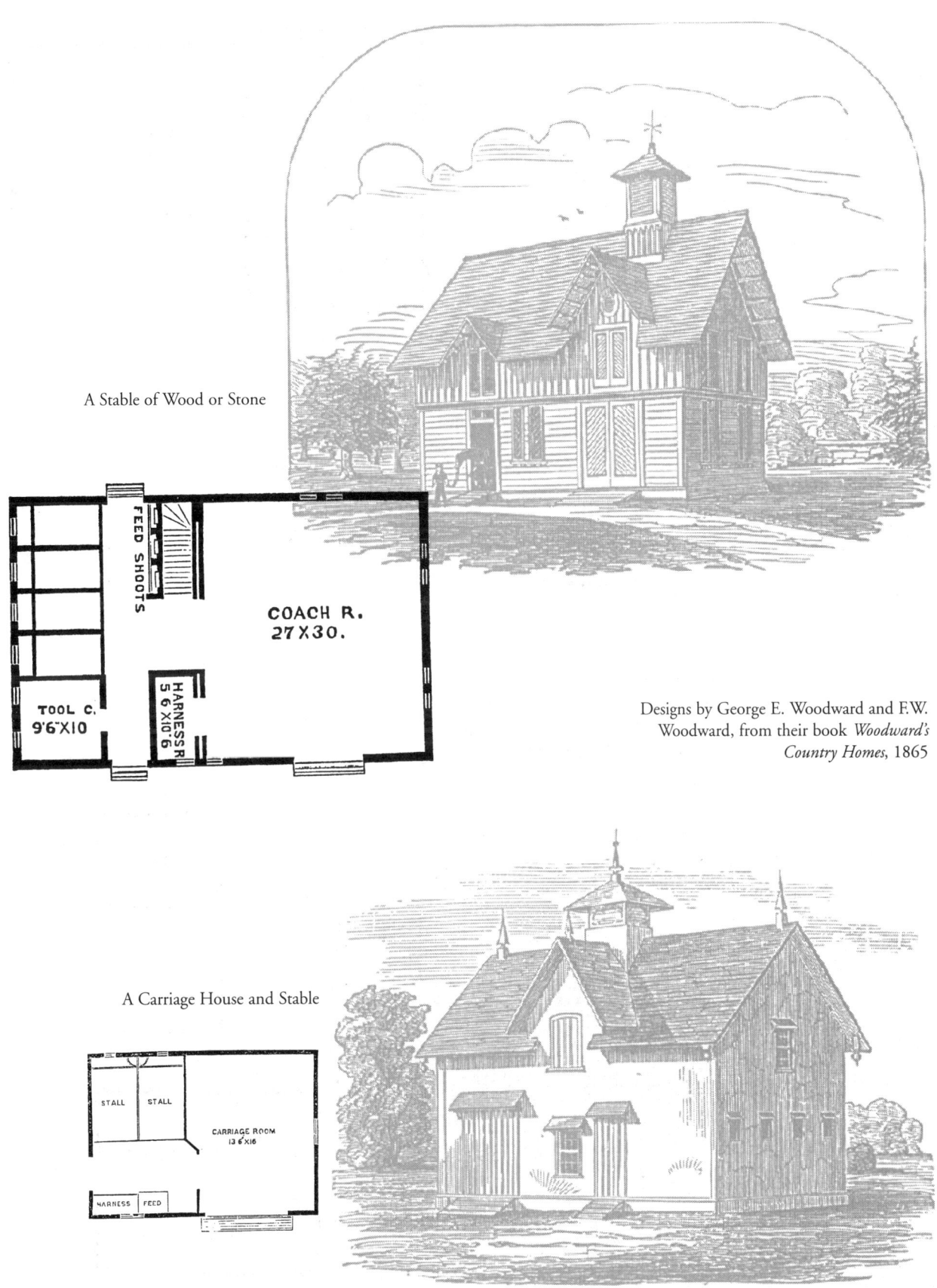

A Stable of Wood or Stone

Designs by George E. Woodward and F.W. Woodward, from their book *Woodward's Country Homes*, 1865

A Carriage House and Stable

Yesterday's Carriage Barns 15

A Stable design, from the 1895 plan catalog, *How to Build, Furnish and Decorate.*

Mail-Order plans for carriage houses were offered between 1883 and the last turn of the century by New York publisher Robert Shoppell and his Cooperative Building Plan Association. They offered blueprints through a popular magazine called *Shoppell's Modern Houses* and a variety of different catalogs.

A Stable design from an 1893 issue of the magazine *Shoppell's Modern Houses.*

A Stable design, from the 1895 plan catalog, *How to Build, Furnish and Decorate*.

A Stable design from an 1893 issue of the magazine *Shoppell's Modern Houses*.

Yesterday's Carriage Barns

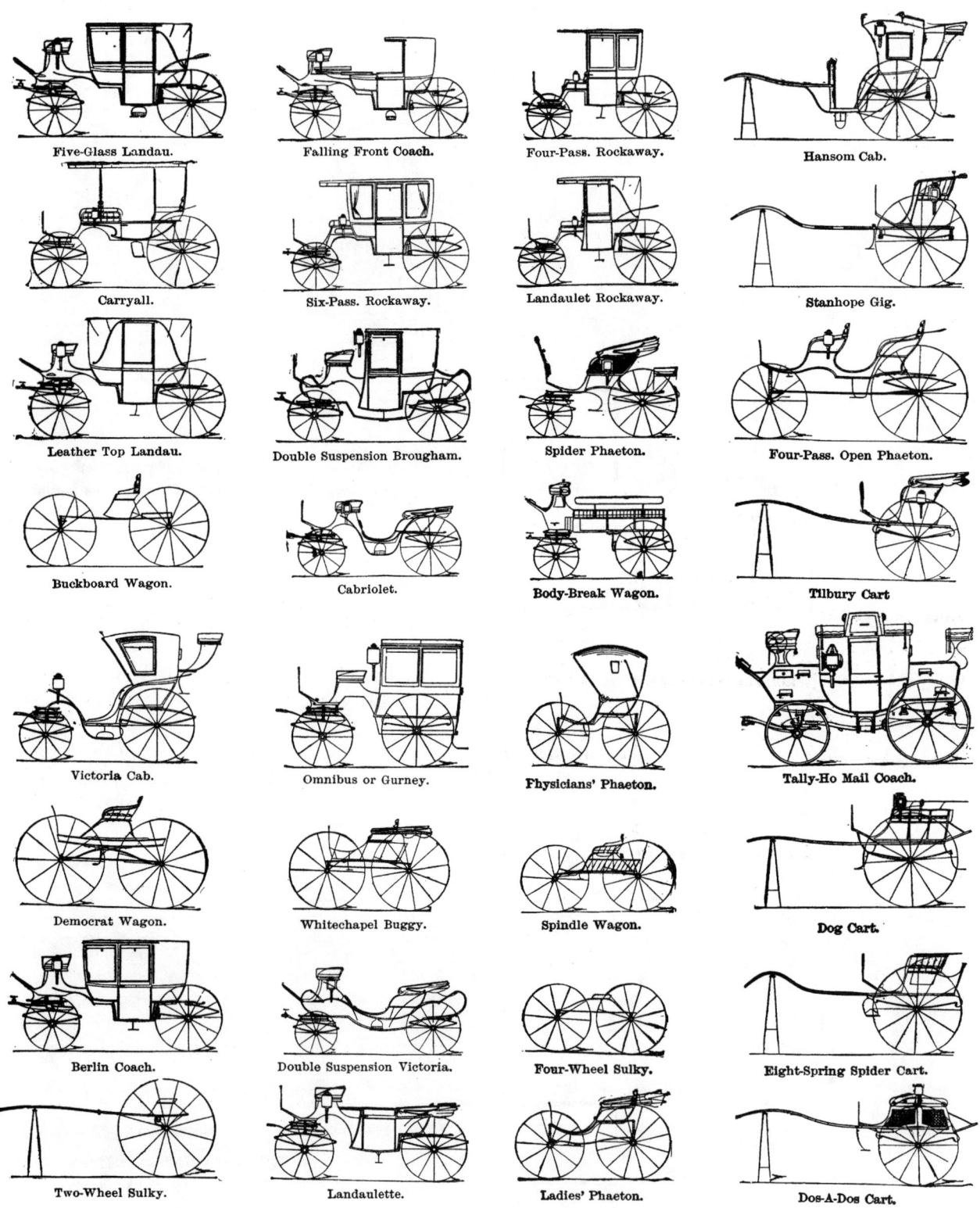

Modern Fashionable Carriages and Vehicles in General Use, from *Hill's Manual of Social and Business Forms*, 1892

18 Yesterday's Carriage Barns

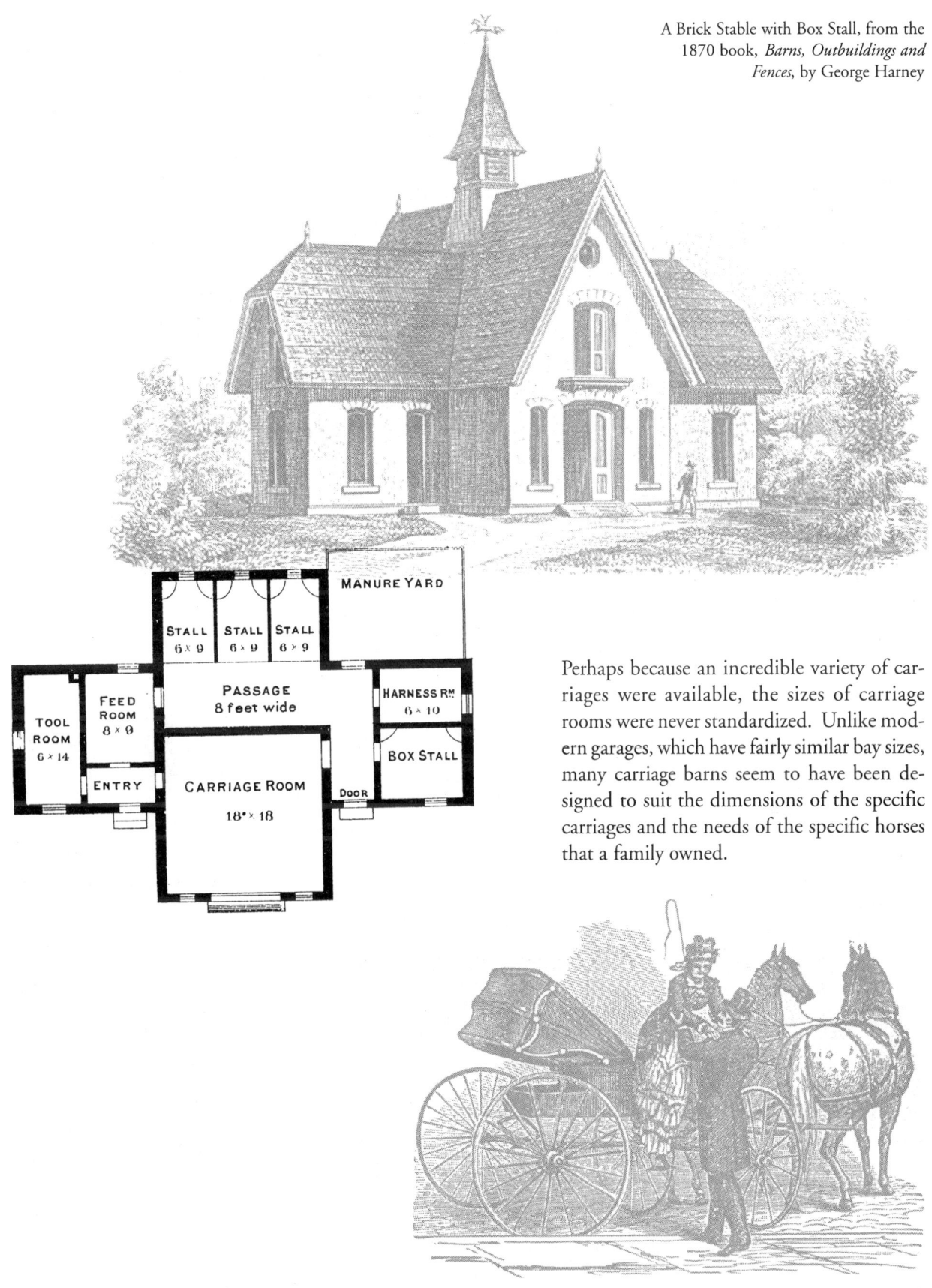

A Brick Stable with Box Stall, from the 1870 book, *Barns, Outbuildings and Fences*, by George Harney

Perhaps because an incredible variety of carriages were available, the sizes of carriage rooms were never standardized. Unlike modern garages, which have fairly similar bay sizes, many carriage barns seem to have been designed to suit the dimensions of the specific carriages and the needs of the specific horses that a family owned.

Yesterday's Carriage Barns   19

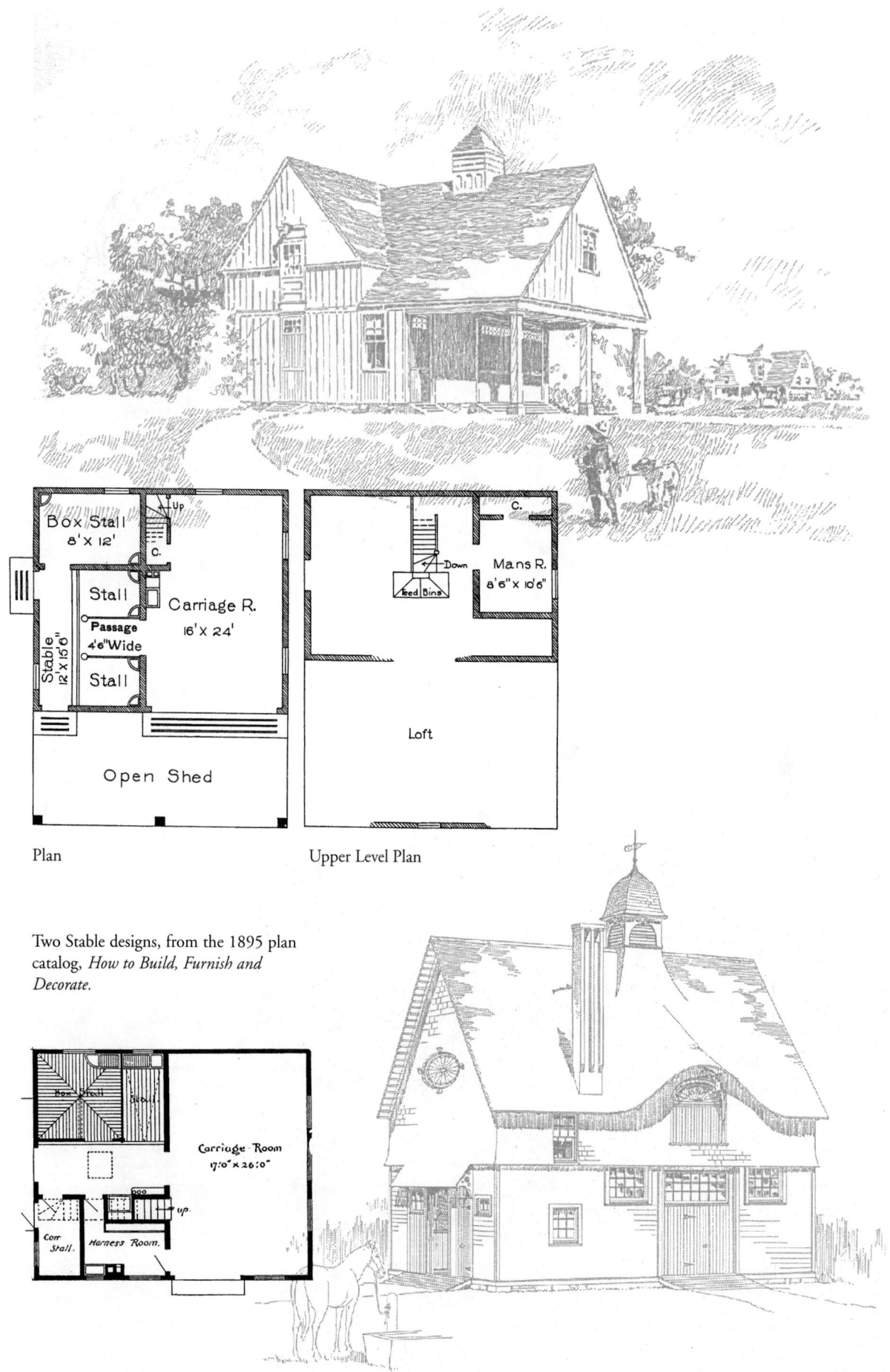

Two Stable designs, from the 1895 plan catalog, *How to Build, Furnish and Decorate.*

20 Yesterday's Carriage Barns

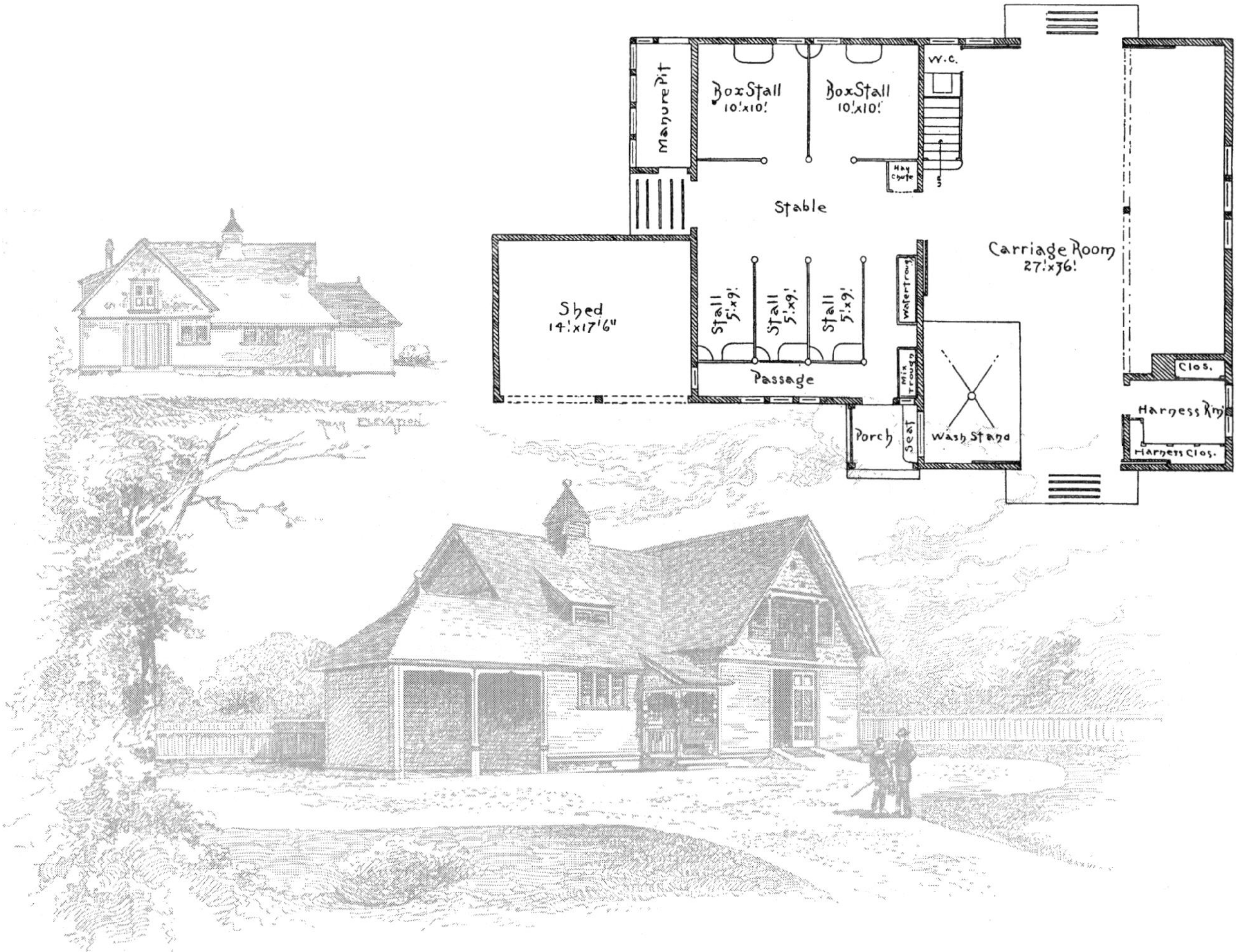

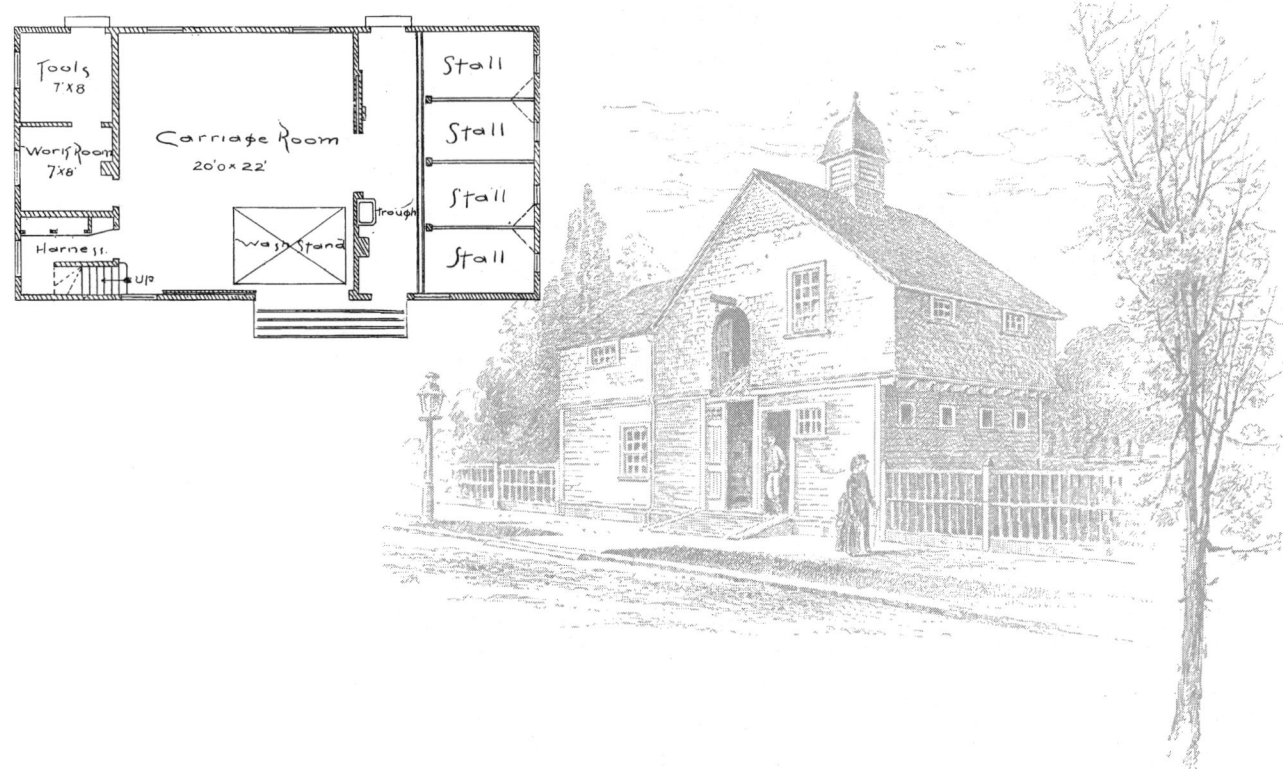

Two Stable designs from 1887 issues of the magazine *Shoppell's Modern Houses*.

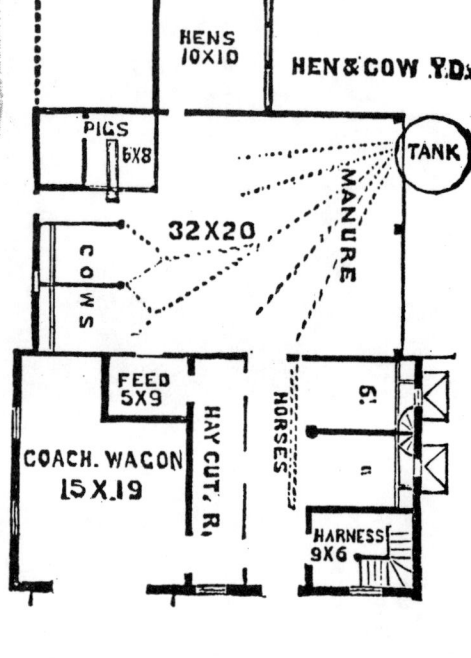

A Barn design, by architects George and F.W. Woodward, From their 1867 book, *Woodward's Architecture, Landscape Gardening, and Rural Art.*

It's fun to look at yesterday's designs for ideas and inspiration, but if you're planning an authentic reproduction, you should also study how carriage barns were placed on properties. The general advice was that carriage barns should be built upwind from the home to protect the family from smells and flies from the stables. On slopped properties, they were built downhill from the home to protect wells and cisterns.

One of the most important elements of building almost any country, village or suburban building in the time of carriage barns is often overlooked in restorations and reproduction designs. It was almost universal for all barns and outbuildings on a property to be built "on the square" with the house. Their walls were parallel or perpendicular to the walls of the house, regardless of how far apart the buildings were. Even when various buildings were different in style, they appear to be part of the same composition if built this way. It's a tradition worth keeping. Build your carriage barn square with your house.

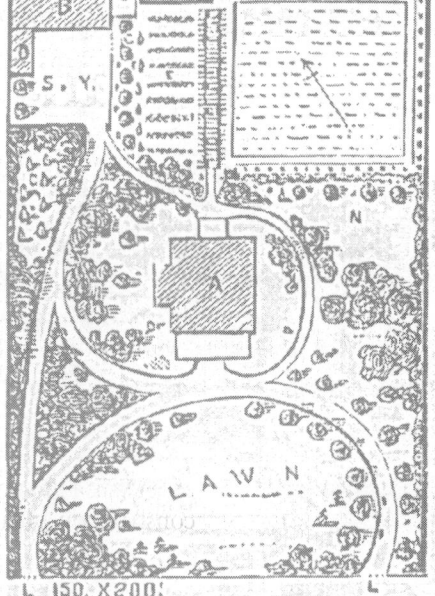

A Landscape Plan, from the 1867 book, *Woodward's Architecture, Landscape Gardening, and Rural Art.*

22 Yesterday's Carriage Barns

Today's Carriage Barns

On the pages of this chapter and in those that follow this one, you'll find some designs that capture a bit of the spirit of yesterday's carriage barns. Most were not planned as reproductions, or to match any specific historical style, but each has something that sets it apart from what most people are building today. You'll find lofts, cupolas, big sliding barn doors, flexible layouts, dramatic roof lines and some unusual shapes to the windows and doors. The designers had fun with these, and if you build one, you will too. Construction plans or easy-to-build kits are available for all of the designs.

When the carriage barn shown above was published in an 1887 issue of the magazine *Shoppell's Modern Houses,* it was estimated to cost $2,200.00 to build in New York City. That price was based on costs that included carpenters' rates of $2.50 per day. Blueprints were sold through the magazine for $30.00.

Today you'll have a hard time finding a good carpenter who will work for $2.50 a day, and your backyard building might cost more than $2,200, but blueprints are still surprisingly inexpensive. Good construction plans will simplify the building process, allow you to get accurate bids from contractors and permits from your building department, and save you the risk of costly misunderstandings during construction. Robert Shoppell, the publisher of the 1887 plan magazine, guaranteed that his blueprints would save builders at least 10% of the cost of a project. The same is true today.

The Almond Grove Barn

768sf Lower Level Space
672sf Storage Loft Space
288sf Each Optional Expansion Shed
Pole Frame Construction

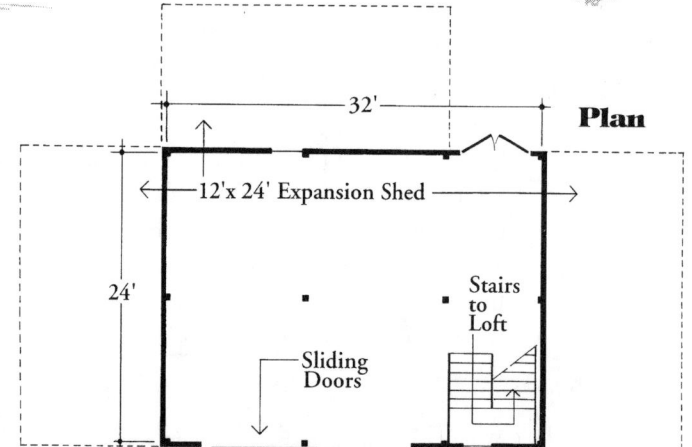

This practical design was originally planned as a small barn for an almond growing operation in California, but it's flexible enough for a great variety of uses. It's designed to take up to three expansion sheds for a total of five full-size parking bays or any combination of parking, workshop area or storage. The main building has wide back doors for yard equipment or a small tractor and stairs to a full storage loft.

Construction plans for The Almond Grove Barn come with drawings for the expansion sheds. Order them through the catalog, *Backyard Barns*, which is available for $4.00 from Donald J. Berg, AIA, PO Box 698, Rockville Centre, NY 11571. You can see this and dozens of other designs on the *Barns, Barns, Barns* website: www.grove.net/~noff/barns.html.

Bethany Coach House

576sf Garage Space
576sf Storage Loft Space
144sf, 216sf or 288sf Optional Expansion Sheds
Pole Frame Construction

The Bethany Coach House with one expansion shed

The Bethany Coach House combines versatility with old-time charm. It's a two-bay garage with big sliding barn doors and stairs up to a full loft. Easy-to-build shed additions can expand it on either side or across the back. The illustration above shows a 12'x24' add-on garage on one side. Smaller additions of 12'x12' and 12'x18' can be used as garden sheds, studios, tractor shelters or workshops. You can use the add-ons to expand your building over the years or to customize it now to suit your needs exactly. Blueprints include drawings for all three sizes of expansion sheds.

For complete information on how to order plans and to see other similar designs, send $4.00 for the catalog, *Backyard Barns*, to Donald J. Berg, AIA, PO Box 698, Rockville Centre, NY 11571 or visit the *Barns, Barns, Barns* website: www.grove.net/~noff/barns.html.

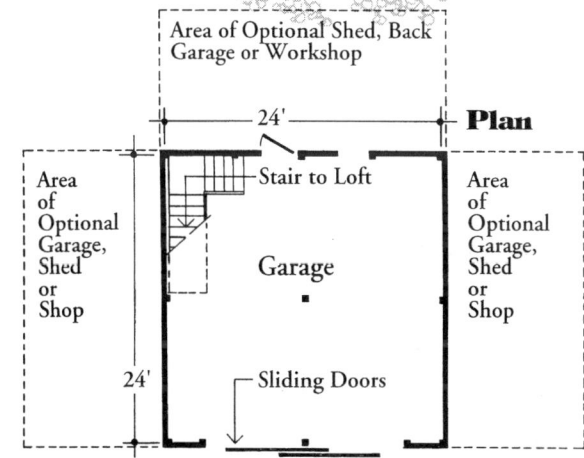

The Bethany Coach House with two expansion sheds

Carriage Barns

Abel's Hill Barn

728sf Garage Space
392sf Storage Loft Space
Conventional Frame Constructions

Copyright, Martha's Vineyard Plans

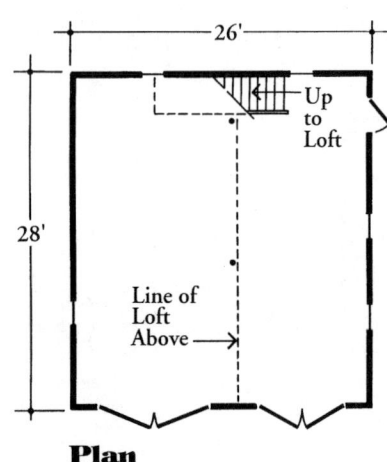

Plan

With a floor plan that's just the size of a modern two-car garage, the Abel's Hill Barn shows the advantage of building an old-style backyard barn. The shed, at one side, is the right size for your car. The bigger bay has a ten foot high ceiling and will fit a boat or camper. Its doors are ten feet wide and nine feet high. Stairs lead to a useful loft that's bright and airy, with windows on all sides.

Compare the versatility of the three different spaces in this building with that of a two-car garage. Now compare the looks. The dramatic height and balanced proportions of this little barn are sure to add value to any country property.

You'll find dozens of attractive small barns, sheds, cottages and homes in the blueprint catalog from Martha's Vineyard Plans. Call 888 847-5267, or send $15.95 plus $3 postage to Martha's Vineyard Plans, PO Box 350, Vineyard Haven, MA 02568. You can also see the designs on a website: www.vineyard.net/biz/mvplans.

All-Purpose Pole Barn

608sf Floor Space
424sf Storage Loft Space
Pole Frame Construction

Copyright, Sheldon Designs, Inc.

You'll find plenty of storage space and a nice, flexible layout in this little pole barn. Use this as a garage and workshop, a crafts barn or as a shelter for tractors or small boats. There is a full set of stairs to a loft. The gambrel roof allows plenty of storage space and 6'-8" loft headroom. Headroom in the main area below is generous at 9'0". Two walk doors and two big sliding doors add to the flexibility of this layout. Because this is a pole barn, it's inexpensive to build and easy to modify to your specific needs.

For information on ordering blueprints and to see other barns, sheds, garages and country homes, order Sheldon Design's catalog. Send $7.00 to Sheldon Designs, Inc., 1330 Route 206 - #204, Skillman, NJ 08588. Or visit their website: www.sheldondesign.com.

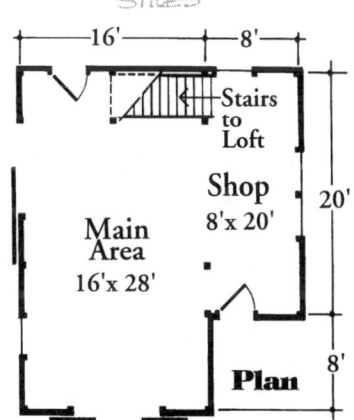

Carriage Barns 27

Waldon Carriage House

1152sf Floor Space
576sf Storage Loft
144sf, 216sf or 288sf Optional Expansion Sheds
Pole Frame Construction

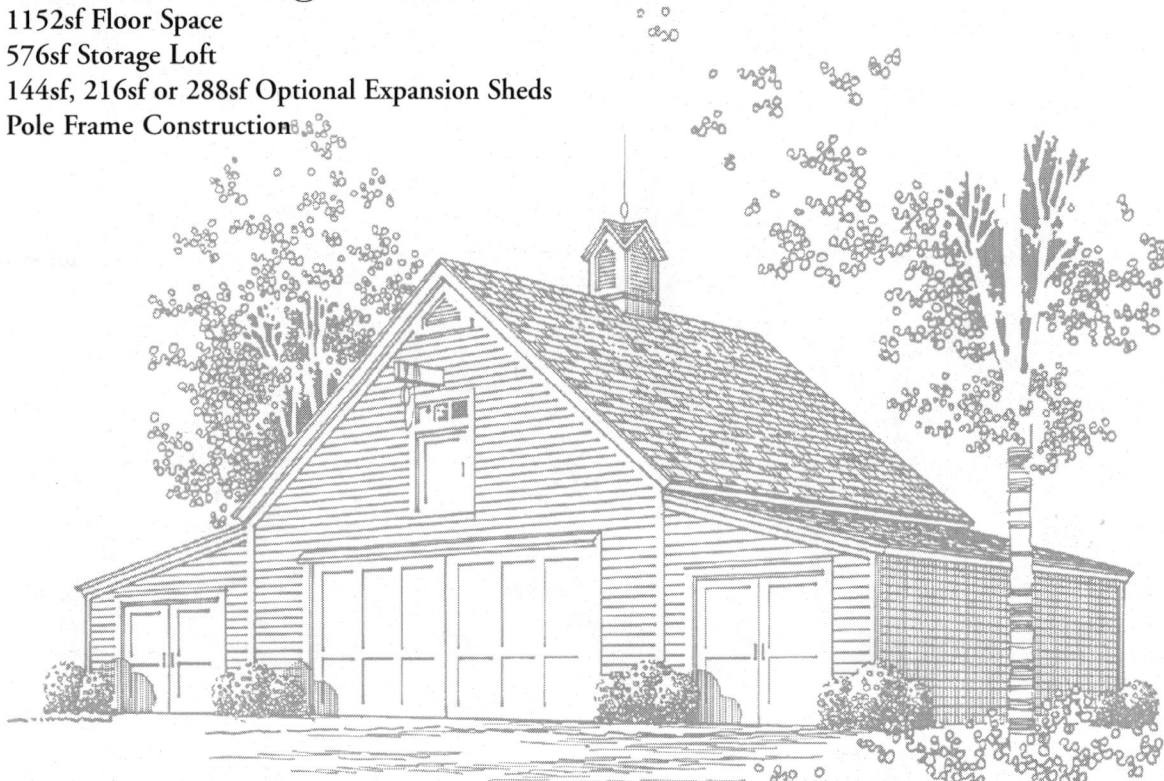

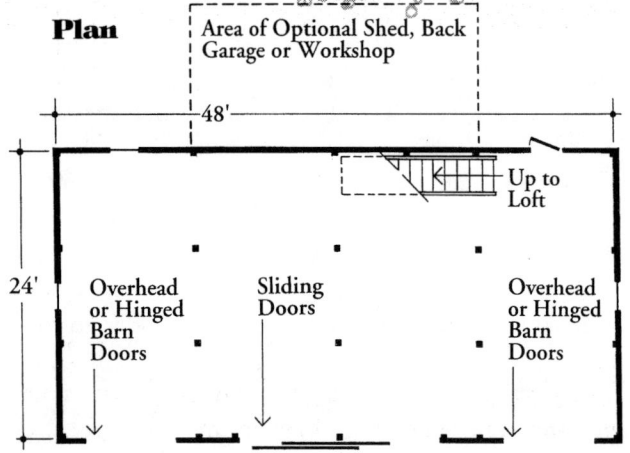

With its pole frame and inexpensive shed-roof wings, the Waldon offers a big area for parking or work space at a very reasonable cost. The center sliding doors are 8' high and 9' wide for two wide RVs, boats or trucks. Two side bays are each as big as a one-car garage. They can have conventional overhead garage doors or hinged barn doors for a carriage house look. An outside hatch door, with a working lift post, and interior stairs give convenient access to a high storage loft. An optional back shed can expand the space to a full 36' deep or serve as a separate garage, tractor shelter or workshop. Blueprints include drawings for three different size back sheds: 12'x12', 12'x18' and 12'x24'.

Send for the $4.00 catalog, *Backyard Barns,* for more information and to see other designs. It's available from Donald J. Berg, AIA, PO Box 698, Rockville Centre, NY 11571. Or, visit the *Barns, Barns, Barns* website: www.grove.net/~noff/barns.html

The Ashford Carriage House

912sf Floor Space
816sf Storage Loft
288sf Optional Expansion Sheds
Pole Frame Construction

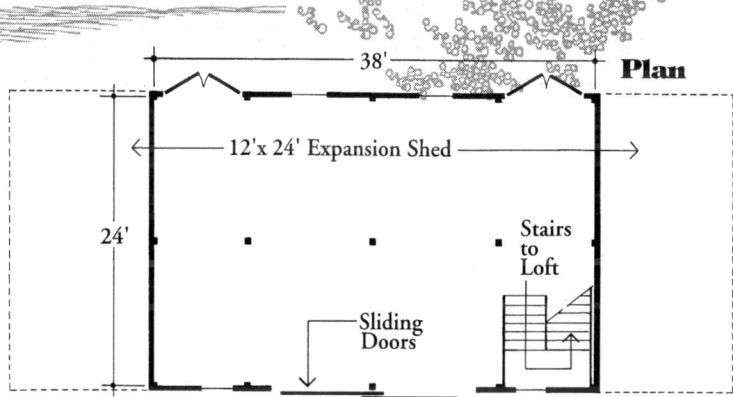

The Ashford Carriage House is a big two-bay garage with added floor space, a walk-up loft and a unique layout. The sliding front doors open 8'-6" high and 9' wide for big SUVs, campers or boats. Two sets of back doors are for tractors, yard equipment, bikes or sport vehicles. Optional expansion sheds provide two more full-size parking spaces or room for a workshop or studio.

The Ashford Carriage House is planned for inexpensive pole frame construction. Blueprints include all the drawings you'll need to build the garage and the expansion sheds. For more information and to see other designs, see the *Barns, Barns, Barns* website: www.grove.net/~noff/barns.html. Or send for the brochure, *Backyard Barns*, for $4.00 from Donald J. Berg, AIA, PO Box 698, Rockville Centre, NY 11571.

The Home Office

550sf Garage
326sf Studio/Office Loft
Conventional Frame Construction

Copyright, Eli Townsend & Son

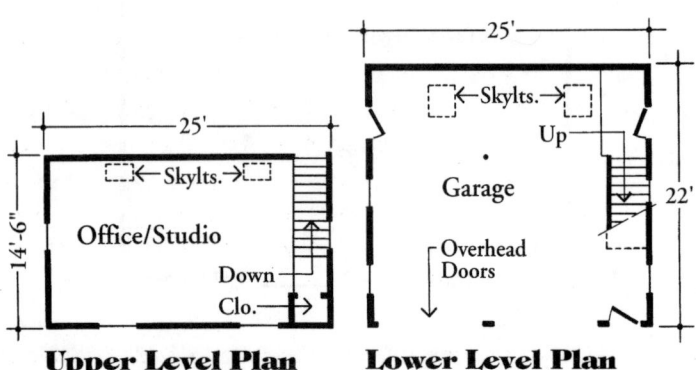

Upper Level Plan   **Lower Level Plan**

This handsome Saltbox style garage, by engineer Harry Townsend, has a big studio space upstairs that's brightly lit by windows and skylights. It's planned as a home office, but it would also be a great space for a game or hobby room or a quiet study away from the hubbub of the house.

For information on construction plans for The Home Office and other New England style garages and cottages, visit the website: http://albino.com/Townsend. Or write for free literature to Eli Townsend & Son, 132 Hemlock Drive, Deep River, CT 06417.

One and A Half Story Saltbox Carriage House

936sf Floor Space
648sf Loft Space
Post & Beam Kit

Copyright, Country Carpenters

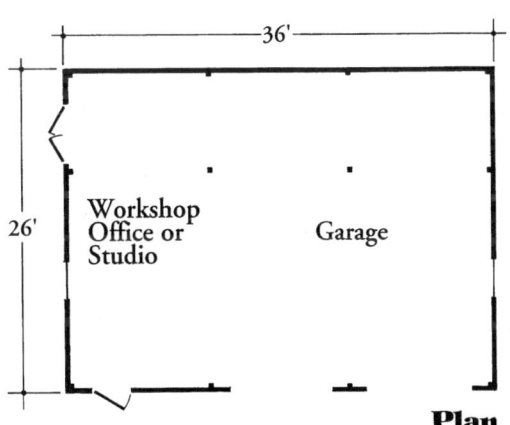

Plan

With its Saltbox style roof and timber frame structure, this building shows its New England heritage. It's a generous 26' deep and can be built at widths of from 24' to 60'. Use it as a garage, workshop, home office, or studio. It's shown here with open parking bays, but it's available with carriage-house doors and a variety of other options. The kit includes a full stairs for access to an 8' high loft.

This is just one of dozens of designs of traditional barns and carriage-houses that are available as pre-cut, easy-to-build kits from Country Carpenters. See their website at www.countrycarpenters.com, or send $4.00 for their catalog to Country Carpenters, 326 Gilead Street, Hebron, CT 06248.

Carriage Barns 31

1860's Carriage Barn

660sf Floor Space
660sf Storage Loft
Conventional Frame Construction

Copyright, Country Designs

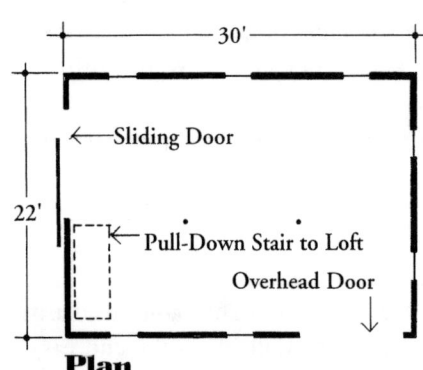

Plan

As you may have noticed from the historic designs in this book, the most common type of carriage house on yesterday's farms and in country villages, were small barns that looked very much like this design. This isn't a reproduction. It's a clever layout for a modern two-car garage that's designed to harmonize with traditional country buildings. The layout has "L" shaped parking spaces with doors on two sides. This works best for the way that most of use our garages - one space for the car and one space for everything else. Have one door open on your driveway and use the other for your lawn tractor, hobby car, boat or whatever. There's generous space in the corner for a workshop or storage and pull-down stairs to a loft.

For information on blueprints and to see dozens of other garages, homes and outbuildings, send $8.00 to Country Designs for their plan catalog. Mail to PO Box 774, Essex, CT 06426.

32 Carriage Barns

Victorian Carriage Barn

660sf Floor Space
660sf Loft Space
Conventional Frame Construction

Copyright, Country Designs

Although the details and proportions of this building make it seem as if it were a hundred years old, the layout makes it work today as a two-car garage. The big sliding doors open on center parking bays. On both sides there are 6'x22' storage or workshop areas. Stairs lead up to a high, usable loft which has lots of sunlight through big windows.

This is just one of the dozens of traditional small buildings offered in Country Designs blueprint catalog. Besides carriage barns, you'll find cottages, sheds, horse barns, workshops, studios, poolhouses, well houses and period fences. For a copy, send $8.00 to Country Designs, PO Box 774, Essex, CT 06426.

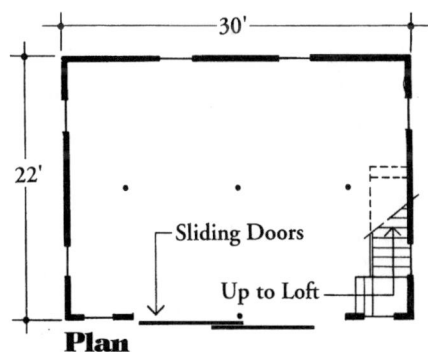

Carriage Barns 33

RV & Car Barn
708sf Floor Space
204sf Loft Space
Conventional Frame Construction

Copyright, Country Designs

This timeless barn design solves a modern problem: where to park one of those huge new SUVs. The hinged barn doors open to 9'-9" wide by 8'-4" high. The space behind them is a full 17' wide by 25' deep. That's big enough for any SUV and for campers, farm machinery and even small boats on their trailers. There is an adjoining space for a one-car garage or shop and a small loft for additional storage.

Country Design's catalog has information on ordering blueprints and designs of many other backyard barns. For a copy, send $8.00 to them at PO Box 774, Essex, CT 06426

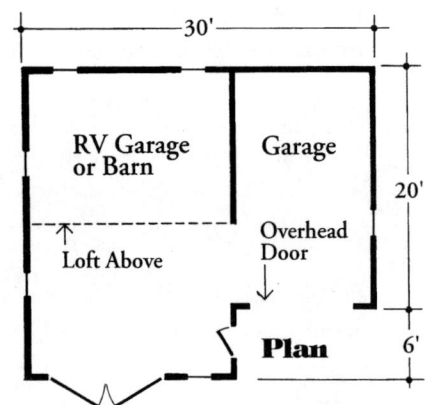

34 Carriage Barns

672sf Floor Space
576sf Loft Space
200sf Each Optional Expansion Shed
Pole Frame Construction

Ashville Barn

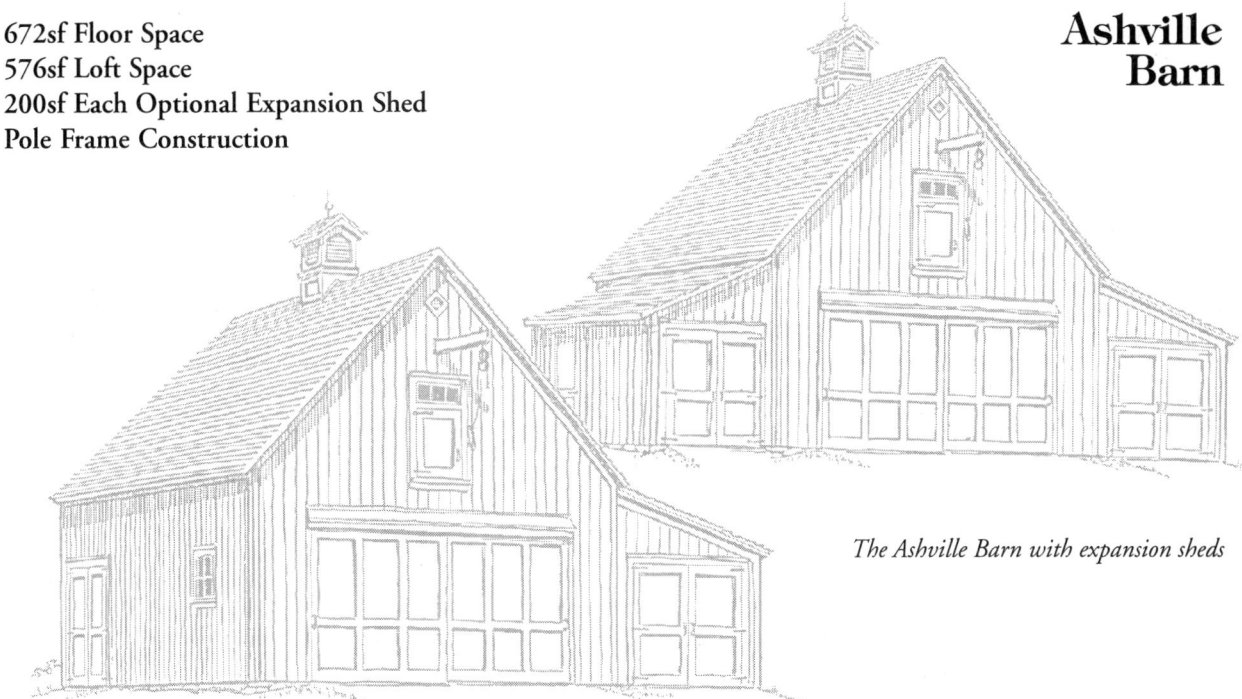

The Ashville Barn with expansion sheds

The Ashville Barn is designed to grow with your needs. Build it as a generous sized two-bay garage with big sliding barn doors and stairs up to a full loft. Then add a 10' x 20' shed on either side for a small car, boat or tractor, or for use as a workshop or garden shed. Need more space? Finish your barn with a second shed for four full bays of storage.

The sliding doors of the main barn are 8'- 6" high and open a full 9' wide on each side, for big SUVs, boats or campers. The main barn also has double hinged doors at back, for easy access for a yard tractor or garden tools. The loft has an outside hatch and a convenient lift-post.

For information on construction plans, send $4.00 for the brochure *Backyard Barns* to Donald J. Berg, AIA, PO Box 698, Rockville Centre, NY 11570. The Ashville is one of dozens of barn designs featured on the new *Barns, Barns, Barns* website. Visit it at www.grove.net/~noff/barns.html

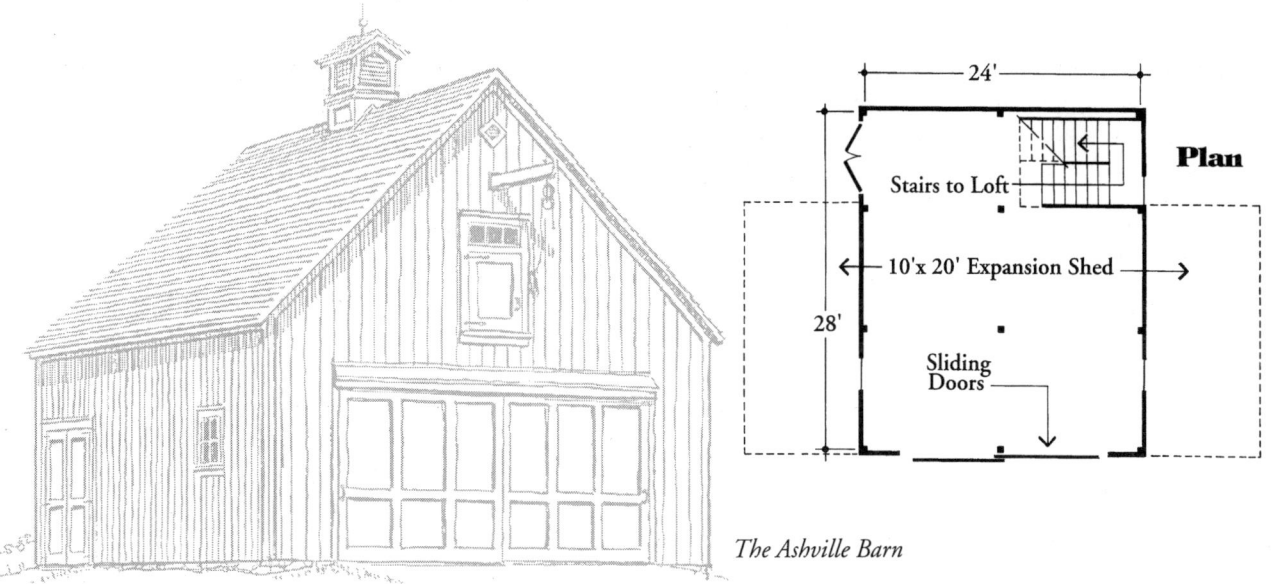

The Ashville Barn

Carriage Barns 35

Concord Barn

296sf Floor Space
208sf Crawl Space Loft
Conventional Frame Construction

The little Concord Barn is perfect if you have limited backyard space. Not much larger than many sheds, it will still fit a small car or tractor with room to spare for a workshop or tools. Or, build it as a two-stall stable or garden shed. Blueprints, from Homestead Design, Inc., include all three of the layouts shown.

Order *The Homestead Design Planbook* to see more than 40 attractive and flexible barns, garages, stables and garden sheds. Send $5.00 to Homestead Design, Inc., PO Box 2010, Port Townsend, WA 98368. You can also see the designs on the Internet at www.homesteaddesign.com.

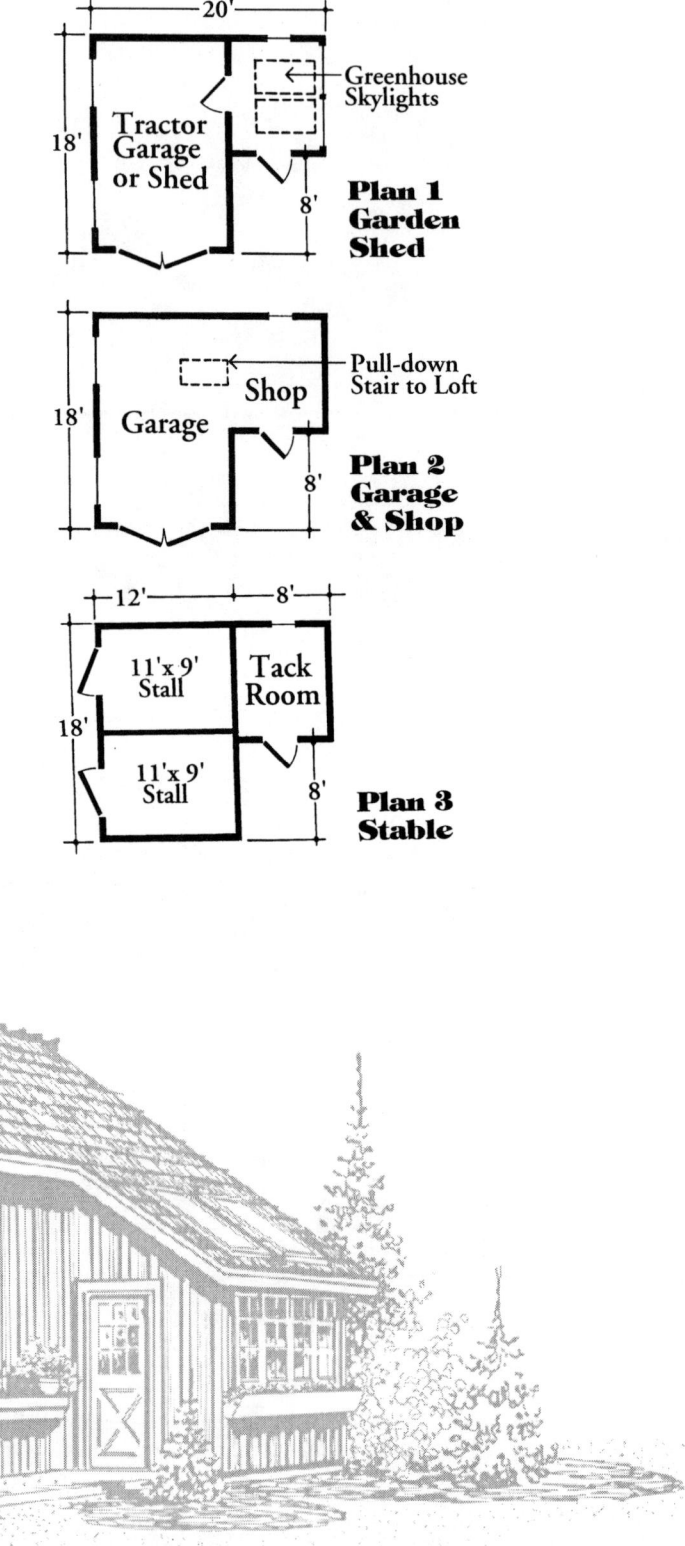

Copyright, Homestead Design, Inc.

36 Carriage Barns

The North Cove

400sf Floor Space
200sf Storage Loft Space
Conventional Frame Construction

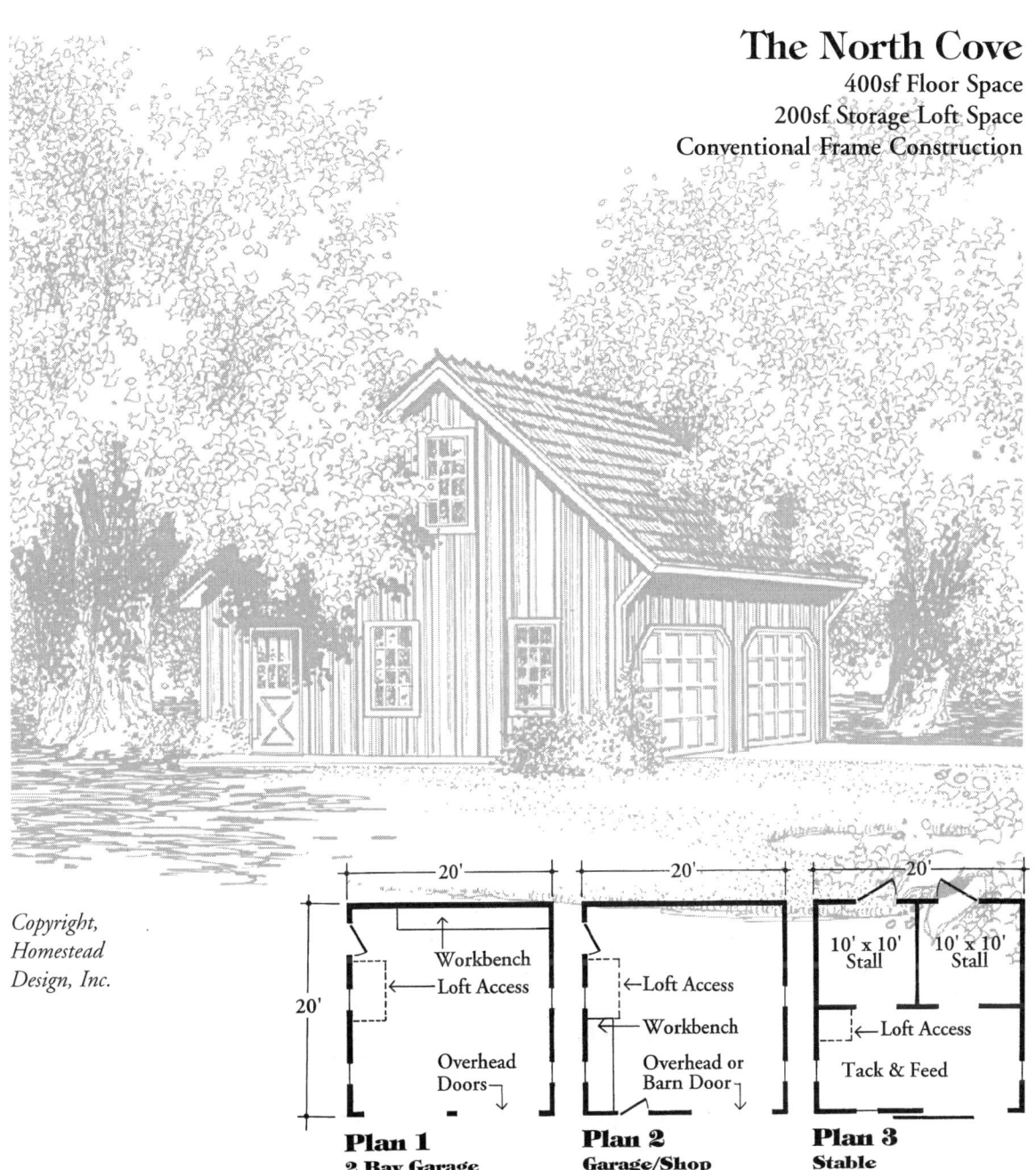

Copyright, Homestead Design, Inc.

Plan 1 — 2 Bay Garage

Plan 2 — Garage/Shop

Plan 3 — Stable

Like many of the country buildings from Homestead Design, the North Cove's blueprints come with a variety of layouts. You can build it as a two-bay garage, as a two-stall stable or as a combination garage and workshop. A storage loft, accessible by ladder or pull-down stairs, adds to the versatility.

You can order blueprints on-line at www.homesteaddesign.com or through their catalog, *The Homestead Design Planbook*. For a copy of that booklet, send $5.00 to Homestead Design, Inc., PO Box 2010, Port Townsend, WA 98368.

You Name It!
384sf - 1,280sf Floor Space
384sf - 1,280sf Loft Space
Conventional Frame and Truss Roof Construction

Copyright, BarnPlans, Inc.

Notice anything unusual about this design? There's no floor plan. That's because BarnPlans, Inc. has a unique service that lets you design the exact carriage barn, stable, garage, barn house or you-name-it that you want. They provide blueprints of the structural shells of seven different size barns: 8'x 12', 12'x 16', 16'x 24', 20'x 30', 24'x 30', 28'x 40' and 32'x 40'. Then, based on your needs, they supply drawings of different components, like the dormers, fancy entryway and barn doors shown above.

The barns all have gambrel roof truss structures. The "barn roof" trusses make a building with generous loft space and no interior walls. You're free to lay out and build the interior partitions exactly where you need them and to plan and place windows and doors where they work best for you. You design your new carriage barn floor plan while **BarnPlans, Inc.** handles the engineering and drafting of the blueprints of the building's shell..

You can only order BarnPlan, Inc.'s construction drawings on the Internet, but their informative website makes it worth your trouble. Visit it at www.barnplans.com.

Stables

A Stable design from 1895 plan catalog *How to Build, Furnish and Decorate.*

The stables on the following pages are just a small sampling of the hundreds of designs that are available as blueprints or building kits from many of the designers and manufacturers featured in this book. If you see buildings that you like in this chapter, or in any of the others, order the catalogs listed or visit the website addresses to see more horse barn designs.

The Gentleman's Horse Barn

720sf - 1,440sf Floor Space
384sf - 768sf Loft Space
Post & Beam Kit

*30' x 36' Horse Barn
with optional details.
Copyright, Country Carpenters*

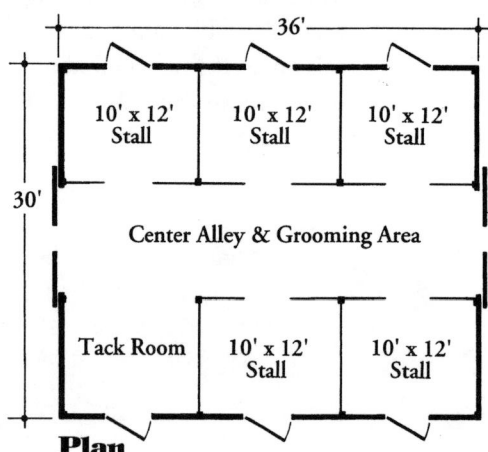

This stable has the look and timber-frame structure of a traditional New England barn. It has 9' high ceilings and a big loft. It's 30' width allows a 10' center aisle and 10' stalls on each side. You can order the length that you need from 24' to 60' and customize your barn with a variety of doors and details. Options include the cupola, Dutch doors, transom window and sliding doors shown plus stairs to the loft, stall grills and more.

Country Carpenters, Inc. will help you design the perfect barn and then ship it to your property as an easy-to-build kit. For complete information and a catalog of other designs, send $4.00 to Country Carpenters, 326 Gilead Street, Hebron, CT 06248. Or, visit their website: www.countrycarpenters.com.

Two-Horse Barn
672sf Barn Floor Space
256sf Loft Space
462sf Garage/Shed Floor Space
Conventional Frame Construction

Copyright, Country Designs

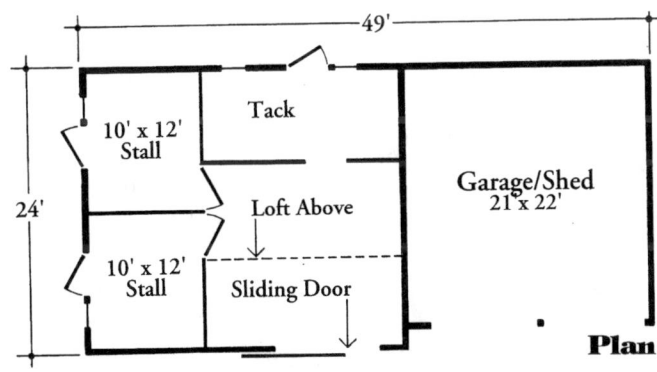

Country Designs captures the look of yesterday's outbuildings in all of their plans. Build this little two-stall stable in your yard and it will seem as if it always there.

The main barn has the stalls, a grooming area and an additional space that can be used for feed and tack or as a small animal pen. An adjacent shed can be built as an open equipment shelter or fitted with doors and used as a two-car garage.

You'll find plans for this design and dozens more in Country Design's blueprint catalog. Send $8.00 to them at PO Box 774, Essex, CT 06426.

Mohawk Barn & Stable

864sf Floor Space
432sf Loft Space
Conventional Frame Construction

Copyright, AshlandBarns

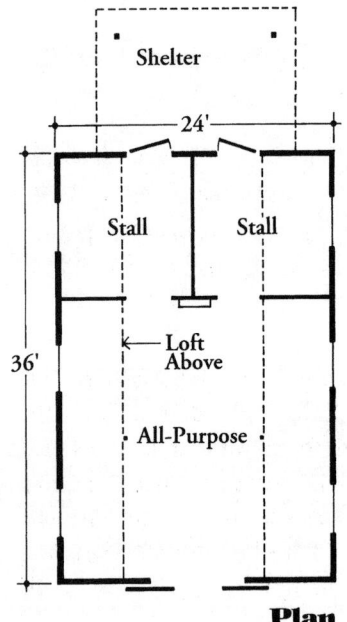

Plan

The Mohawk barn was planned as a combination stable and garage. You could also use it as a workshop or equipment shed. Without the stalls, two cars or full size tractors could stack for parking or repairs. There's a ladder and exterior hatch to a full-length loft with clearstory windows that light and cool the whole building.

For information on how to order blueprints of the Mohawk and to see a catalog of dozens of other barn designs, send $5.00 to AshlandBarns, 990-BH Butler Creek Road, Ashland, OR 97520. AshlandBarns also offers an attractive line of cast aluminum weather vanes, like the one shown on this barn. To order their windvanes and post signs catalog, send $2.00 to the same address.

Yakima Horse Barn

864sf Floor Space
432sf Loft Space
Conventional Frame Construction

Copyright, AshlandBarns

The Yakima barn has three 12' x 12' stalls, another 12' x 12' area for tack and feed, a pass-through center alley, a large shady shelter, and a big storage loft. You can reach the loft by an interior ladder or through a convenient exterior hatch. The monitor-style roof allows high windows for light and ventilation.

If you like the look and practicality of the Yakima barn but want a different layout, you're in luck. AshlandBarns offers inexpensive blueprints for a dozen more monitor-style barns plus a great variety of gambrel roof barns, garages, offices, country homes, sheds and workshops. You can see the designs on their website: www.ashlandbarns.com. Or, order their $5.00 catalog by mailing to AshlandBarns, 990-BH Butler Creek Road, Ashland, OR 97520.

The Cambridge Barn & Stable
864sf Floor Space
864sf Loft Storage Space
Conventional Frame Construction

Copyright, Homestead Design, Inc.

The gambrel roof of this little barn creates a loft that has a full 8'-6" of headroom. The loft doubles the usable space in this building. It has convenient sliding exterior access doors at both ends.

Like many of Homestead Design's barns, a clever structural design with just two structural posts on the main floor gives you a variety of options with the floor plan. You can build the Cambridge with one, two or three stalls. Substitute a loft ladder for the walk-up stairs and you can squeeze in another pony. Or, leave areas open as a big workshop or equipment storage area.

You can see dozens of other barns and stables on Homestead Design's new website: www.homesteaddesign.com. Or, order their catalog of blueprints, *The Homestesd Design Planbook*, by sending $5.00 to Homestead Design, Inc., PO Box 2010, Port Townsend, WA 98368.

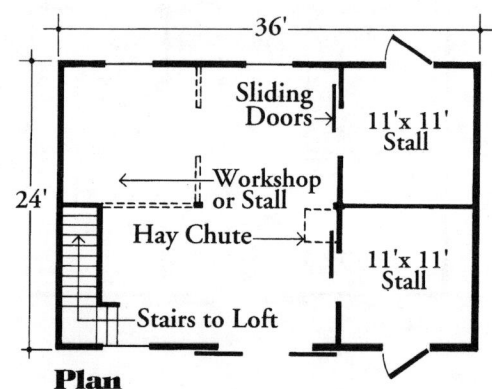

44 Stables

Live-Ins

Only the well-to-do had help to drive their carriages and care for their horses. Yet, almost all 19th century carriage barns had "groom's rooms" upstairs. For farm families, this extra room was used for farm hands hired to help with the harvest. For others, it was a rental, a store room or a place for guests to stay. The fact that carriage barns had bedrooms or small apartments seems to have inspired the common use of the term "carriage house" for any garage or barn with living space upstairs.

Here's an assortment of live-in barns and garages. You can use any of them as your residence, a rental unit, a guest house, a vacation home or as a backyard studio or office.

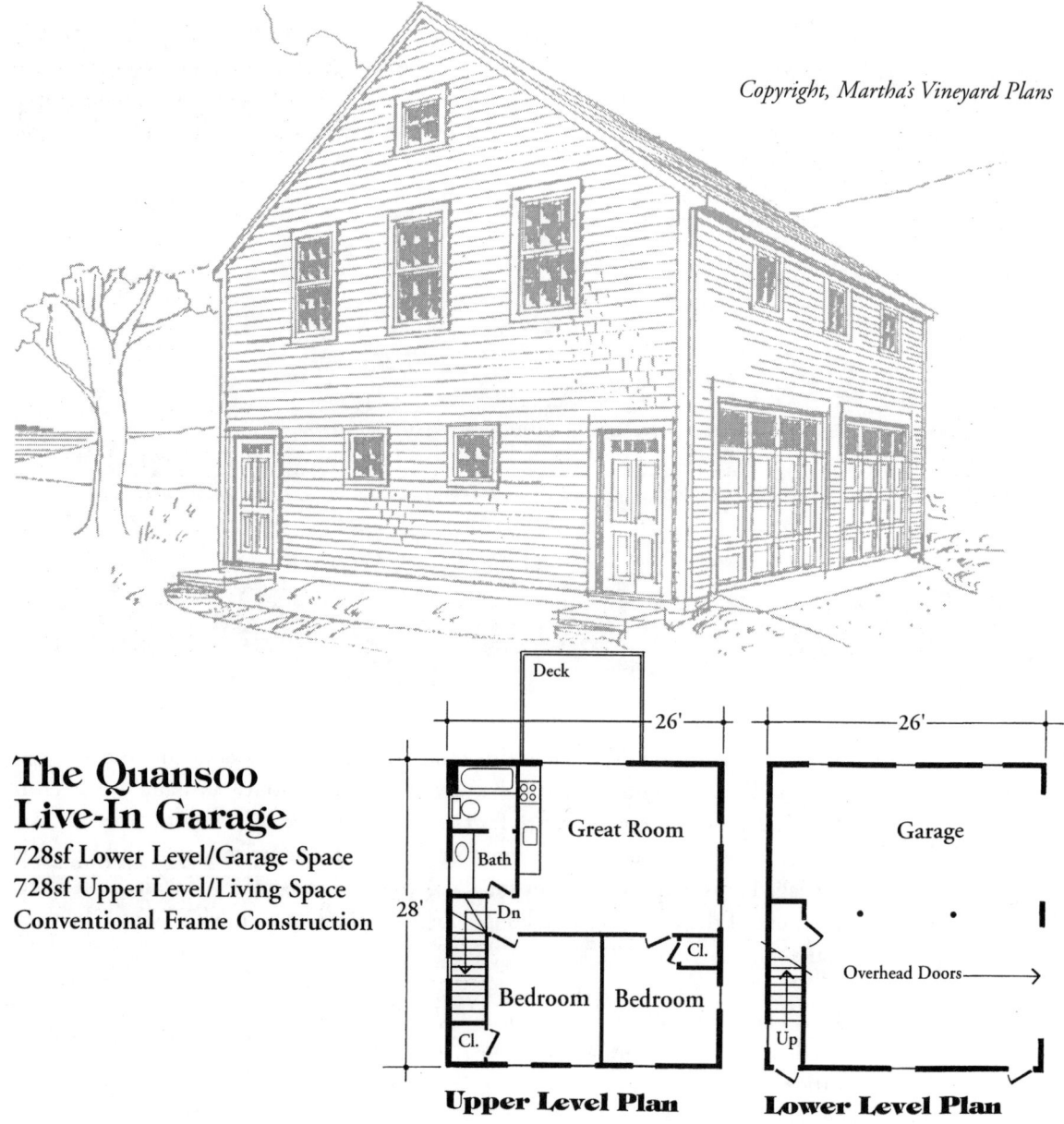

Copyright, Martha's Vineyard Plans

The Quansoo Live-In Garage

728sf Lower Level/Garage Space
728sf Upper Level/Living Space
Conventional Frame Construction

Like many of the live-in designs that you'll see in this chapter, the Quansoo will make a great starter home, a rental unit, a backyard office, a guest house or an apartment for any family member who wants a little independence. It has the simple good looks of a New England barn and even a bit of the Shaker style in its balance of windows and doors. Only the exterior is Spartan. The upper level of this building is a complete little home with two bedrooms, a full bath, an all-purpose room and glass doors to a big deck.

For more information on blueprints for the Quansoo, order the planbook from Martha's Vineyard Plans, PO Box 350, Vineyard Haven, MA 02586. It's available for $15.95 and $3 postage and includes designs for seventy barns, garages, sheds, cottages and homes. You can also see the designs on-line at www.vineyard.net/biz/mvplans.

Hickory Studio
880sf Lower Level Garage Space
659sf Upper Level Living Space
Conventional Frame Construction

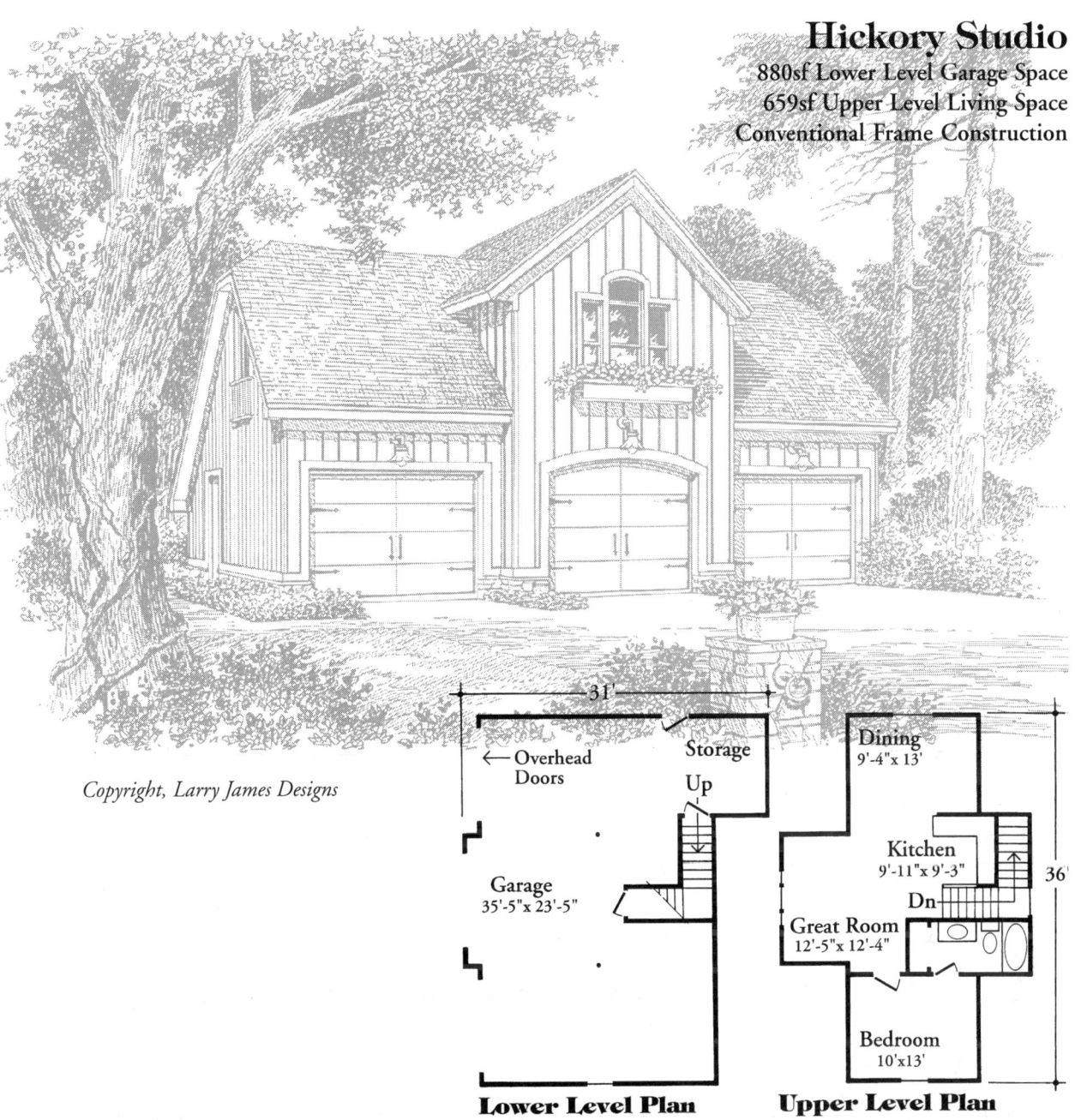

Copyright, Larry James Designs

Lower Level Plan **Upper Level Plan**

The Hickory Studio is a three-bay garage with a comfortable studio apartment upstairs. Its carriage house proportions and details make it the perfect match for last century's homes and new classic-revival style houses.

Complete construction plans are available from Larry James Designs, Monroe, Louisiana. Call 800-742-6672 for information. The Hickory Studio and other traditional home designs can be seen on a website. Visit www.larryjames.com.

Live-Ins 47

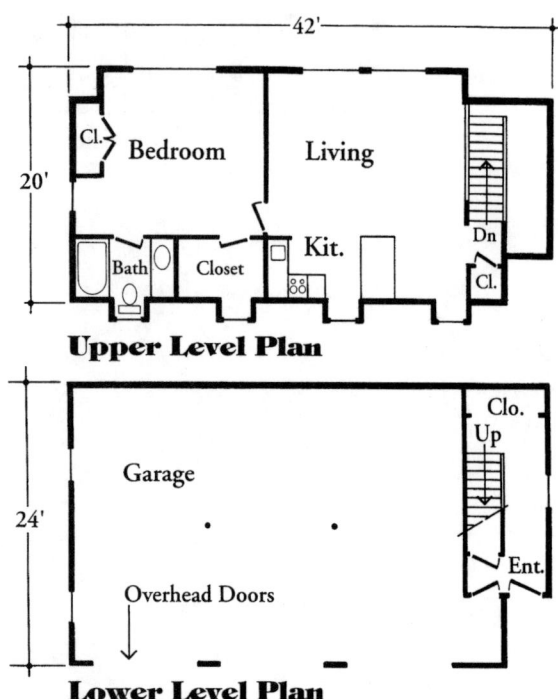

Upper Level Plan

Lower Level Plan

This gracious garage design features an attractive, separate entry to an upstairs apartment. Dogshed dormers on the front and a full dormer across the back allow plenty of windows for a bright, comfortable home up there.

You can order Sheldon Designs' complete catalog for information on blueprints for this design. It includes designs fore more than 60 garages, studios, barns, cabins, cottages, sheds and utility buildings. Send $7.00 to Sheldon Designs, Inc., 1330 Route 206, #204, Skillman, NJ 08558. You can also visit the website: www.sheldondesigns.com or see many of the designs on the Barn Blueprint website: www.grove.net/~noff/barns.html.

Three-Car Garage & Studio Apartment
851sf Lower Level Garage Space
759sf Upper Level Living Space
Conventional Frame Construction

Copyright, Sheldon Designs, Inc.

Two-Car Garage & Studio Apartment

590sf Lower Level Garage Space
510sf Upper Level Living Space
Conventional Frame Construction

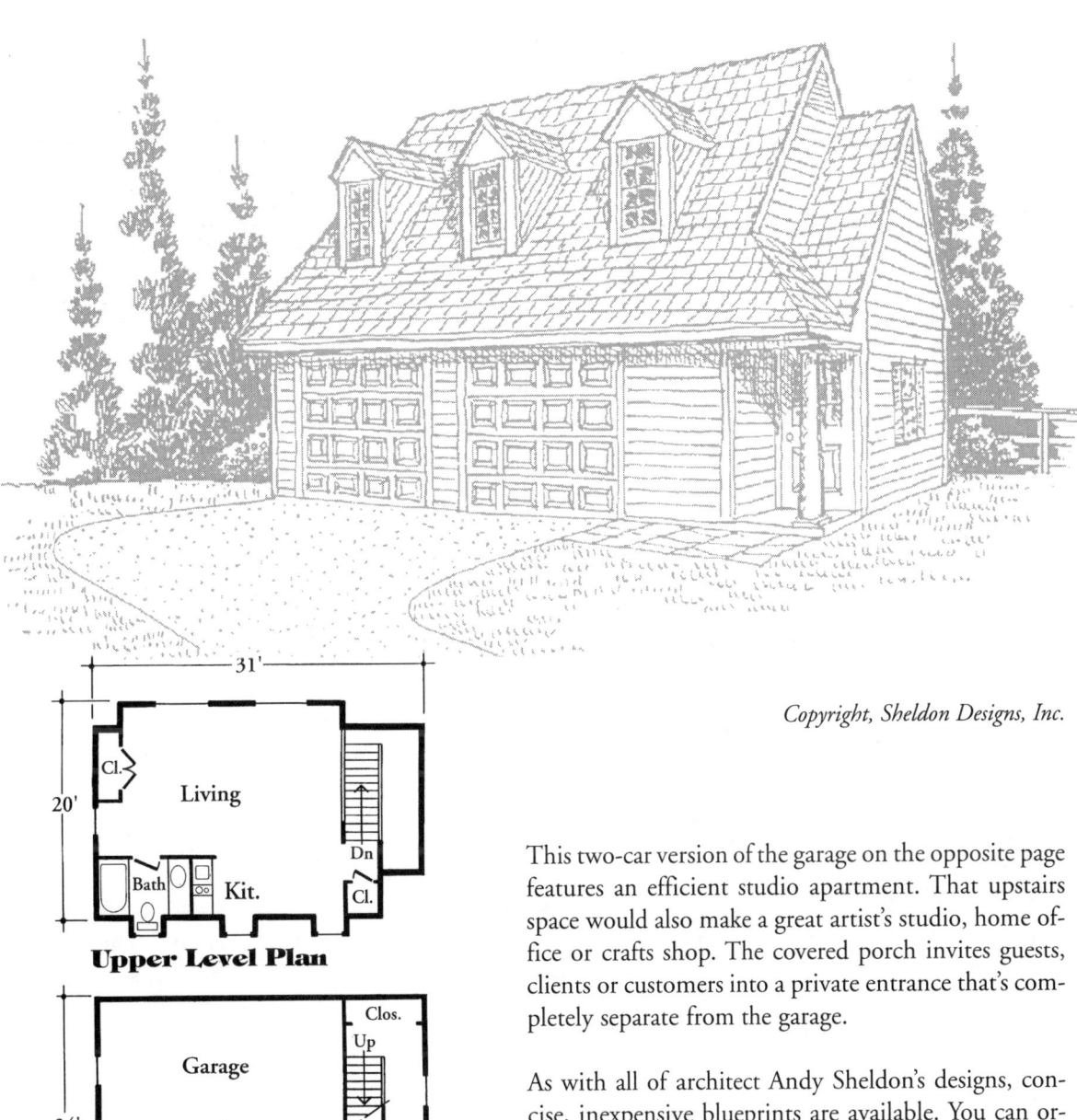

Copyright, Sheldon Designs, Inc.

This two-car version of the garage on the opposite page features an efficient studio apartment. That upstairs space would also make a great artist's studio, home office or crafts shop. The covered porch invites guests, clients or customers into a private entrance that's completely separate from the garage.

As with all of architect Andy Sheldon's designs, concise, inexpensive blueprints are available. You can order them through his $7.00 plan catalog. Write to Sheldon Designs, Inc., 1330 Route 206, #204, Skillman, NJ 08558. Or visit the website: www.sheldondesign.com.

Live-Ins 49

The Rockport Barn
740sf Ground Floor Space
676sf Loft Living Space
Conventional Frame Construction

Copyright, McKie Wing Roth, Jr.

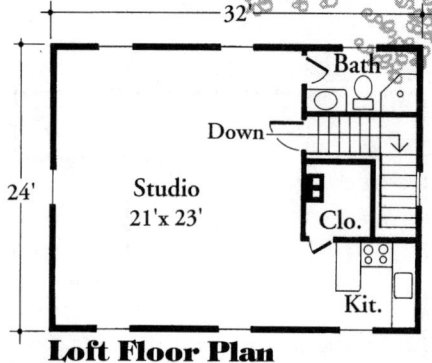

Loft Floor Plan

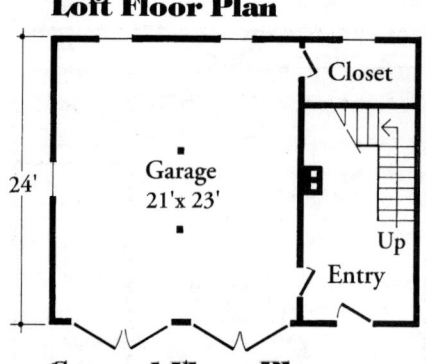

Ground Floor Plan

The Rockport Barn is an elegant, traditional design with a variety of modern uses. The lower level is a full two-bay garage with a large storage room at back for tools and yard equipment. Stairs, in a separate entry, lead to a bright, 21' x 23' room, a full bath, a small kitchen and a big storage closet. You can use the loft as a studio, home office, guest cottage or apartment.

For information on ordering blueprints, order the catalog, *New England Style Home Designs*, for $18.00 from McKie Roth Design, Inc., PO Box 31, Castine, ME 04421, call 800-232-7684, or visit the website: www.mckieroth.com.

50 Live-Ins

The Castine Barn

834sf Ground Floor Space
770sf Loft Living Space
Conventional Frame Construction

Copyright, McKie Wing Roth, Jr.

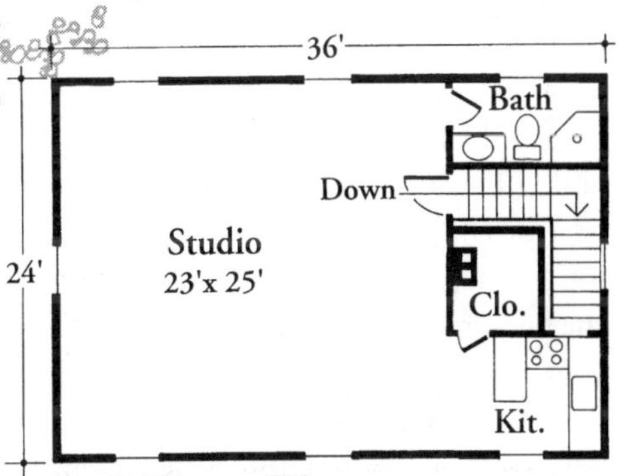

McKie Roth is known for creating new homes that capture the spirit of New England traditions. He built the Castine Barn to look like an old Maine barn, but also to serve as his own efficient design studio. He offers complete blueprints so that you can reproduce it as a garage and home office, as a caretakers' apartment or rental, or as a guest house.

The Castine Barn is one of many homes and backyard barns presented in the plan portfolio called *New England Style Home Designs*. You can order the catalog for $18.00 from McKie Roth Design, Inc., PO Box 31, Castine, ME 04421, or by calling 800-232-7684. You can also see many of Roth's designs on his website: www.mckieroth.com.

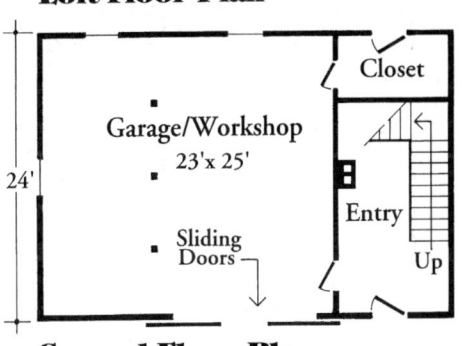

Loft Floor Plan

Ground Floor Plan

Live-Ins 51

This 24' x 32' Horse Barn is typical of the finished look of Shelter-Kit's barns.

Two-Car Parking Barn & Studio Apartment
576sf Lower Level Garage
288sf Upper Level Living Space
Post & Beam Kit

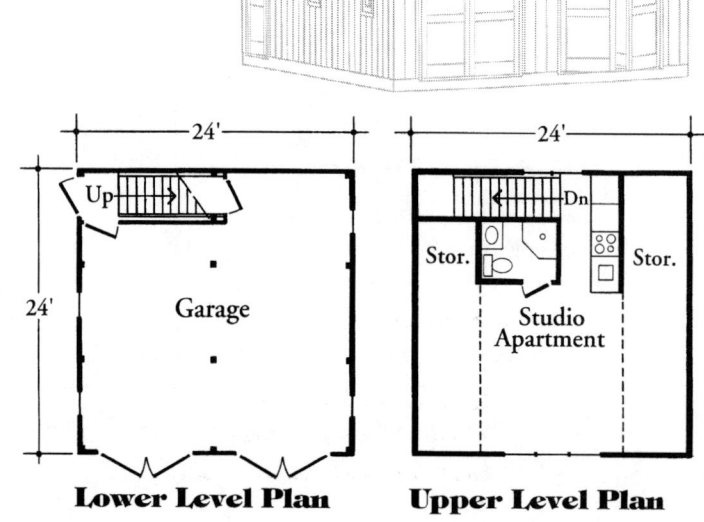

Shelter-Kit, Inc. has a very untraditional way of offering you traditional New England barns. They provide you with blank plans of their standard barns, by mail or from their website, and ask that you sketch out your ideas. Then they provide you with a computer aided drawing of your design. If it's just what you want, they will manufacture an easy-to-build kit and ship it to your site, with a step by step building instruction manual. The shell of a 24'x 24' barn, like the design shown here, can be put together by amateur builders in about two weeks. Of course, for the kitchen and bath, you'll probably need the help of an experienced plumber and electrician.

All of Shelter-Kits' barns are post and beam structures, like some 300 year old barns that are still standing in New England. So, your garage should hold up pretty well. They can build barns, garages and stables in standard widths of 16' and 24' and in lengths of from 16' through 64.'

See Shelter Kit's website for complete information. It's at www.shelter-kit.com. Or, call 603 286-7611 or write for free literature to Shelter Kit, Inc., 22 Mill Street, Tilton, NH, 03276.

52 Live-Ins

Double Duty

568sf Lower Level Garage Space
542sf Upper Level Living Space
Conventional Frame Construction

Copyright, Eli Townsend & Son

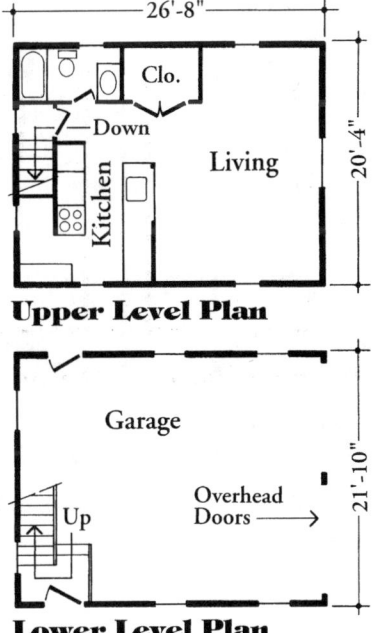

The Double Duty looks as much like a traditional New England gambrel roof cottage as it does a garage. This is a fine little building that could serve as a home, guest house, rental or home office.

Construction plans include drawings for a matching two floor cottage. For free literature, write to Eli Townsend & Son, 132 Hemlock Drive, Deep River, CT 06417. You can see this and other traditional cottages and garages on a website. Visit http://albino.com/Townsend.

Live-Ins 53

Gambrel Garage & Apartment
712sf Lower Level Space
617sf Upper Level Living Space
Conventional Frame Construction

Copyright, B.W. Smith

The lower level of this little building combines a two-car garage, storage and utility rooms and a gracious entry for the upstairs apartment. On the second floor, there's a complete one-bedroom home with generous sized rooms and plenty of windows and closets.

This design is just one of dozens of creative adaptations of classic New England buildings from CADSmith Studio of Brookline, NH. You can see their cottages, barns, homes and garages on their website: www.cadsmith.com. You'll also find information on the site about ordering complete construction plans of any of the designs.

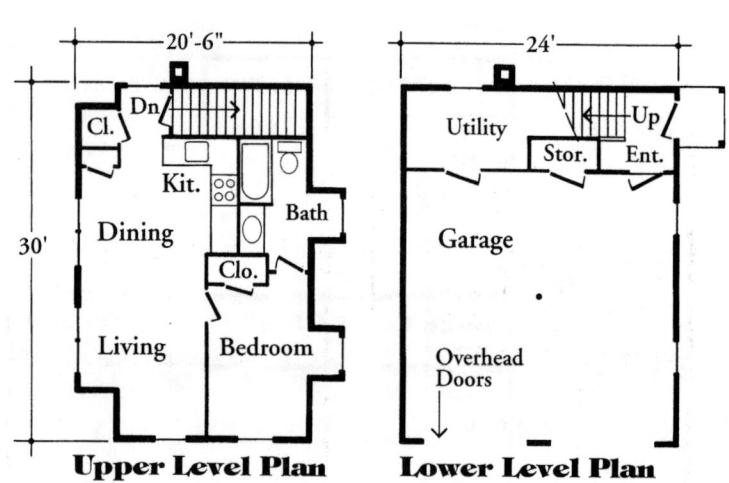

Three-Bay Garage & Apartment

1064sf Lower Level Space
936sf Upper Level Living Space
Conventional Frame Construction

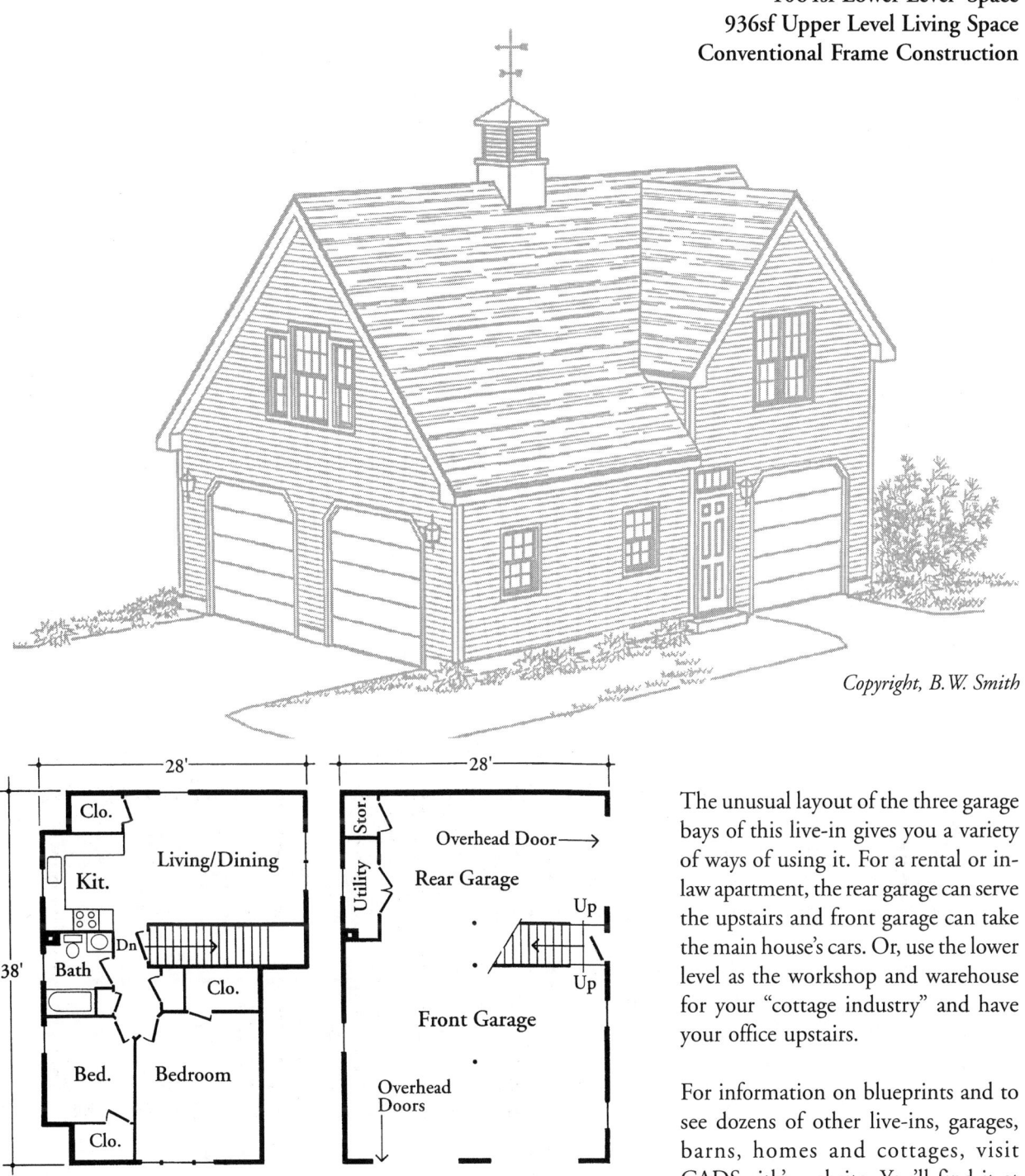

Copyright, B.W. Smith

Upper Level Plan

Lower Level Plan

The unusual layout of the three garage bays of this live-in gives you a variety of ways of using it. For a rental or in-law apartment, the rear garage can serve the upstairs and front garage can take the main house's cars. Or, use the lower level as the workshop and warehouse for your "cottage industry" and have your office upstairs.

For information on blueprints and to see dozens of other live-ins, garages, barns, homes and cottages, visit CADSmith's website. You'll find it at www.cadsmith.com.

Live-Ins 55

Garage Apartment

816sf Lower Level Space
782sf Upper Level Living Space
Conventional Frame Construction

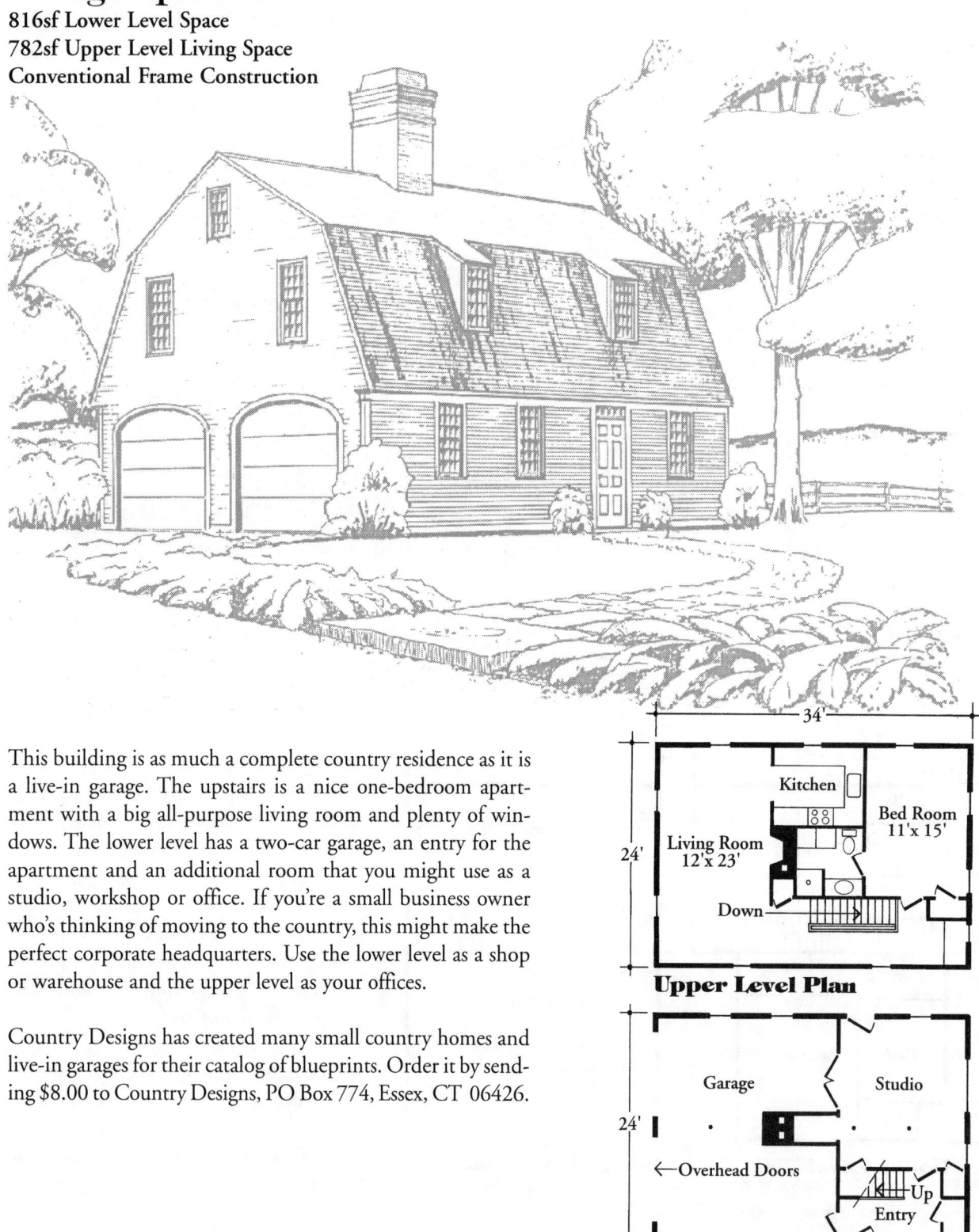

This building is as much a complete country residence as it is a live-in garage. The upstairs is a nice one-bedroom apartment with a big all-purpose living room and plenty of windows. The lower level has a two-car garage, an entry for the apartment and an additional room that you might use as a studio, workshop or office. If you're a small business owner who's thinking of moving to the country, this might make the perfect corporate headquarters. Use the lower level as a shop or warehouse and the upper level as your offices.

Country Designs has created many small country homes and live-in garages for their catalog of blueprints. Order it by sending $8.00 to Country Designs, PO Box 774, Essex, CT 06426.

56 Live-Ins

Loft Garages

All old carriage barns had big lofts for hay. Today, if you need a garage, there are still some benefits to building a loft.

Lofts make garages look more dramatic. Loft garages are usually better proportioned than ordinary low-roofed garages. But, it's the loft itself that's the big advantage. Loft space is very inexpensive. You'll have to build a roof and a foundation for your garage anyway. Add an extra floor between them and you can double your storage space at a cost that's usually less than 25% more. Loft space is high and dry. It's perfect for clothes, family records, photos and other things that you'd worry about keeping down below. Or, use the space as a game room, studio or hobby shop. Build a loft and you'll find a way to fill it.

Sierra Garage

576sf Floor Space
336sf Loft Space
Conventional Frame Construction

Copyright, Homestead Design, Inc.

The Sierra Garage has a saltbox-style roof. It's lower in the front, and that makes the garage seem smaller than it really is. This is a great design if you need a loft for storage space but don't want your new garage to seem out-of-scale with the other garages in a residential neighborhood.

There's plenty of room for two cars and a workbench on the first level and stairs up to a useful 24' x 14' storage loft. A small back porch is a shady place for garden projects.

The Sierra is one of dozens of garages, stables, sheds and small barns that you'll find in the *Homestead Design Planbook*. Order that catalog of blueprints for $5.00 from Homestead Design, Inc., PO Box 2010, Port Townsend, WA 98268. Or see the designs on-line at www.homesteaddesign.com.

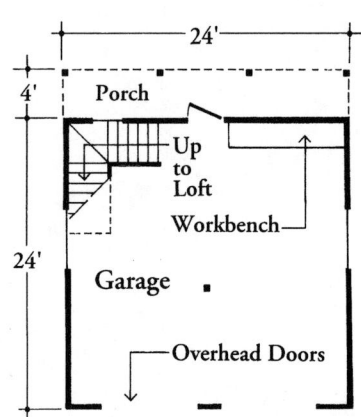

58 Loft Garages

Kensington Three Bay Garage

936sf Floor Space
525sf Loft Space
Conventional Frame Construction

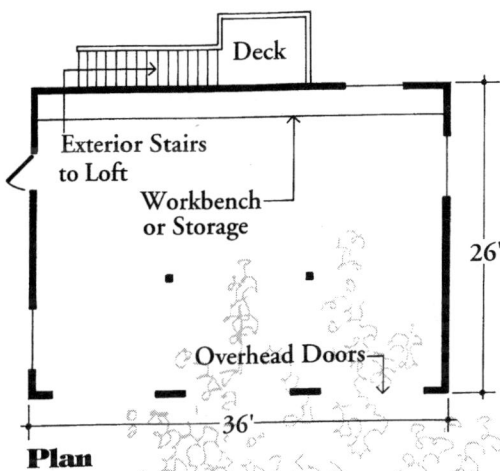

The Kensington has space for three big cars, a workbench and plenty of storage. An exterior staircase leads to a loft with 9' headroom. Like the Sierra, this is a big garage that looks smaller from the front because of its saltbox style roof.

For information on ordering blueprints, visit Homestead Design's website: www.homesteaddesign.com. Or, order their plan catalog by sending $5.00 to Homestead Design, Inc., PO Box 2010, Port Townsend, WA 98268.

Copyright, Homestead Design, Inc.

Two-Bay Country Garage

768sf Floor Space
448sf Loft Space
Conventional Frame Construction

Copyright, Behm Design

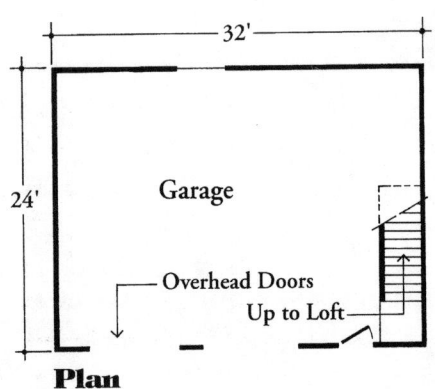

Plan

This pretty and practical country garage has features that set it apart from the rest. There's a high and bright second floor loft for storage or for your use as a studio or office. The loft floor is framed with plywood truss-joists. They make support columns unnecessary, so the entire lower level is free of interior obstructions.

Information on ordering blueprints of this and dozens of other garages is available on-line at www.behmdesign.com. Or, order *Garage Plans: A Catalog of Mail Order Plans* by sending $12.00 to Behm Design, 23632 Highway 99, Suite F-112, Edmonds, WA 98026.

Three-Bay Country Garage

960sf Floor Space
480sf Loft Space
Conventional Frame Construction

Copyright, Behm Design

Like the Two-Bay Garage on the previous page, this big garage has an open plan with no interior columns, room for a work bench below the stairs and a full loft with plenty of windows for light and ventilation.

Well-engineered construction blueprints for this and 67 other garages and storage buildings are available from Behm Design. The drawings are guaranteed to meet your building department's approval under all normal circumstances. Visit www.behmdesign.com to see samples or to download a complete catalog. Or, call 800-210-6776 to order a catalog or plans.

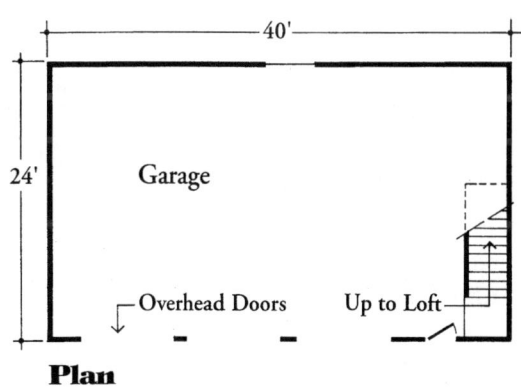

Three-Bay Garage
1139sf Lower Level Floor Space
511sf Loft Storage Space
Conventional Frame Construction

Copyright, BGS Plan Company

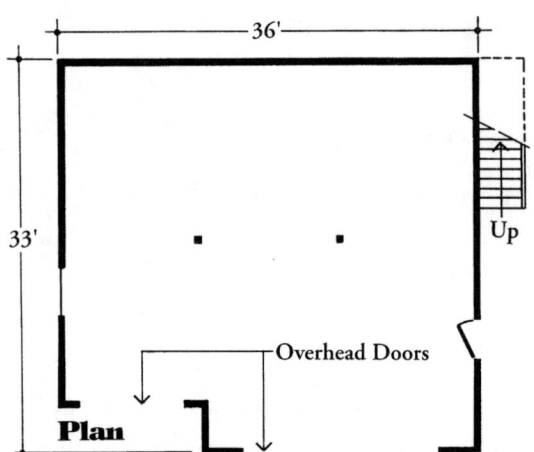

If you can't find the perfect garage in this book, you'll do well to see BGS Plan Company's catalog or website. They offer an amazing selection of well engineered buildings. The three-car garage shown here is just one of more than 1,200 of their garages, barns, stables, workshops and utility buildings. Their blueprints are the most complete you'll find. Full-size patterns, accurate details and 3D framing isometrics that are keyed to a complete materials list, make the construction easier.

To order the catalog, send $10.00 and $4 postage to BGS Plan Company, PO Box 1181, Roseville, CA 95678. The website can be found at www.bgsplanco.com.

Parking Barns

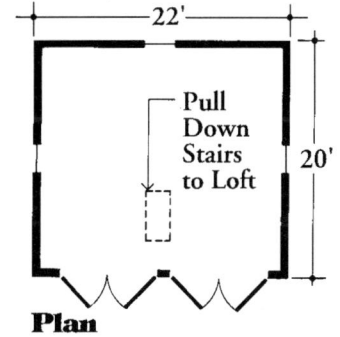

Stockbridge Barn and Stonetree Buggy Barn Combined. Copyright, McKie Roth Design

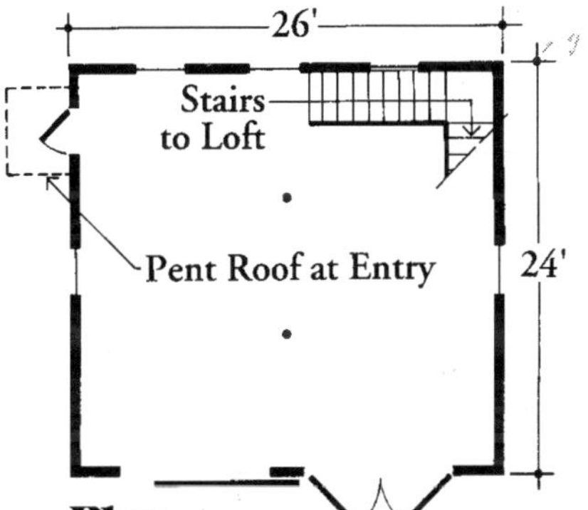

Stockbridge Barn
624sf Lower Level Garage Space
588sf Storage Loft Space
Conventional Frame Construction

These two garages by Maine designer McKie Roth have the details and proportions of traditional New England barns. They'll look great behind any country home. Lofts add to the storage space and convenience of both. These are from Roth's plan catalog which includes small barns, cottages and homes. Order *New England Style Home Designs* by sending $18.00 to McKie Roth Design, Inc., PO Box 31, Castine, ME 04421. You can also see the designs on a website: www.mckieroth.com.

The illustration above shows a way of building a big garage with the look of a rambling, old-time barn. Roth's garages have similar details, so they look great together. Build the Stockbridge and Stonetree barns next to each other and you'll have spaces for four cars or a big workshop, plenty of loft space, and a building that looks perfect in a country setting.

Stonetree Buggy Barn
440sf Lower Level Garage Space
210sf Storage Loft Space
Conventional Frame Construction

Loft Garages 63

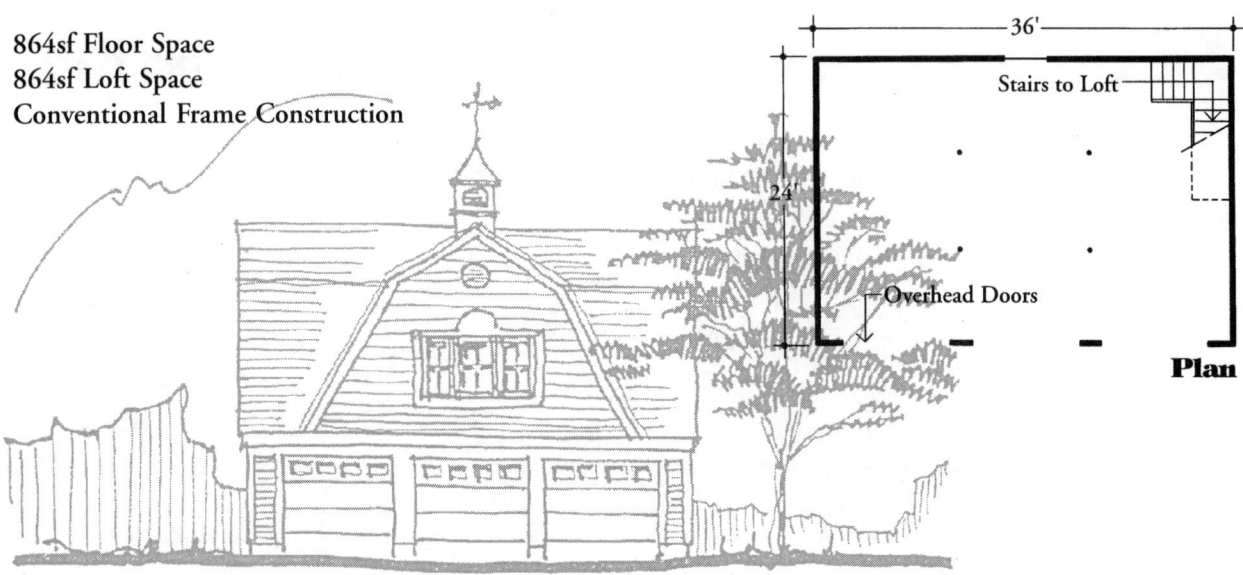

864sf Floor Space
864sf Loft Space
Conventional Frame Construction

Three-Bay Gambrel Garage

These big garages would make graceful additions to the yards behind traditional classic-revival style homes. Their efficient plans and full-size lofts make them work as well as they look. Use these as garages, workshops, studios or home offices.

Just Outbuildings offers carefully detailed construction drawings for both of these designs and for a variety of other garages, sheds, studios and garden buildings. See them on-line at www.justoutbuildings.com, or send $6.00 for a catalog to Just Outbuildings, PO Box 42, Brewster, NY 10509.

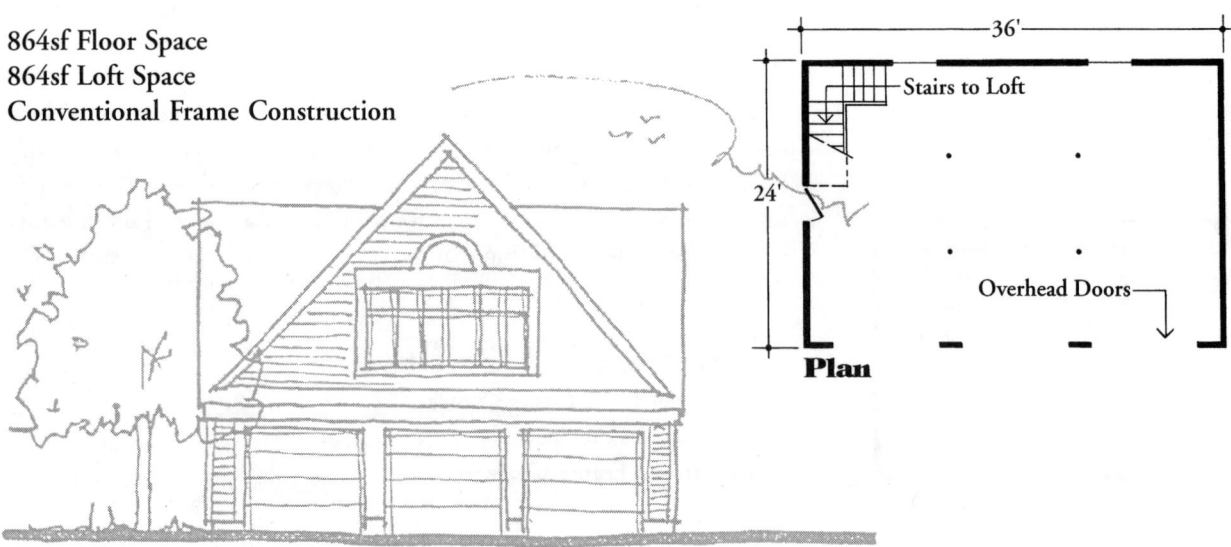

864sf Floor Space
864sf Loft Space
Conventional Frame Construction

Classic Three-Bay Garage

64 Loft Garages

Gambrel Garage
576sf Floor Space
370sf Loft Space
Conventional Frame Construction

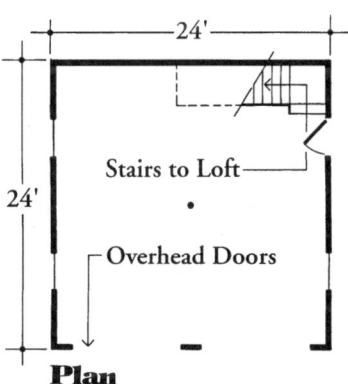

This elegant two bay garage would grace a country estate or suburban street. The steeply pitched gambrel roof ends in gently curved "flying eaves" at the bottom. Clapboard siding and classic details add to the charm. The steep roof allows a high loft that would make a great office or studio. Windows at both ends and dormer windows on both sides brighten the space.

Just Outbuildings offers carefully detailed construction drawings of a variety of garages, sheds, studios and garden buildings. You can see them on-line at www.justoutbuildings.com. Or, order their $6.00 catalog by writing to Just Outbuildings, PO Box 42, Brewster, NY 10509.

Cupolas

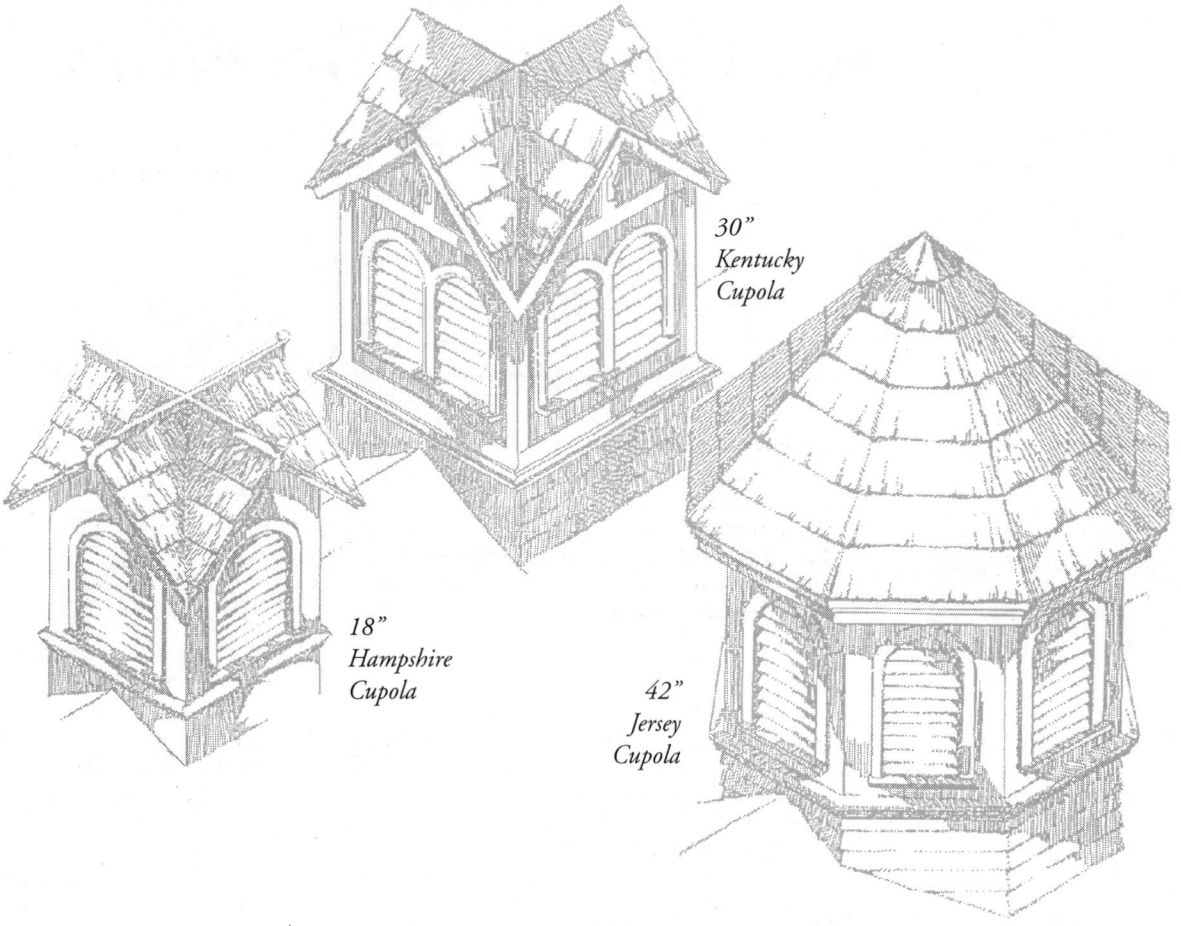

30"
Kentucky Cupola

18"
Hampshire Cupola

42"
Jersey Cupola

Many of the buildings in this book would look great topped with a cupola. A traditional cupola is the perfect base for a weather vane and is still a good way to ventilate a workshop or stable.

You'll find a number of cupola manufacturers in the Directory on the following pages. If you'd like to try to build one yourself, start with a good set of blueprints, like the ones available from H.T.Cadd & Blueprint Service. The ones shown above are just a few of the dozens of cupolas, ventilators, bell towers and ridge-top bird houses that they offer. The construction plans are concise, easy-to-read and surprisingly inexpensive. You can see their cupolas on a website: www.htcadd.com. Or, send $4.00 for the *Cupola Blueprints* catalog to H.T. Cadd & Blueprint Service, N8939 Townline Road, East Troy, WI 53120.

The most common mistake in building a cupola today is building it too small. Traditional cupolas were much bigger than those usually used now. The simplest rule-of-thumb is to select a design that's at least one inch wide for every foot of length of the building's ridge. As an example, if you're building a one-car garage with an 16' long ridge, the Hampshire Cupola, shown above, would be right. For a 40' wide barn, you'd need something the size of the Jersey Cupola.

The Barn Building & Restoration Directory

The manufacturers, designers and services listed here submitted their information for your perusal.

Acorn Forged Iron

Acorn Forged Iron
457 School Street, P.O. Box 31, Mansfield, MA 02048
Decorative builders' hardware for exterior and interior doors, cabinets, gates and shutters. Catalog: $10.00. Free product literature. Phone: 800 835-0121. Fax: 800 372-2676. E-mail: indo@acornmfg.com. Website: www.acornmfg.com.
New Forged Hardware

Allen Cupolas
2242 Bethel Road, Lansdale, PA 19446
Allen Cupolas has a complete selection of cupolas in poplar, redwood and cedar, with copper, brass or aluminum roofs. They have custom sawmill capabilities on premises. Free brochure. Phone: 610 584-8100 or phone, then Fax: 215 699-8100. E-mail: vickallen@msn.com
Cupolas, Woodwork

Annapolis Weathervanes
107 Summers Run, Annapolis, MD 21401
Hundreds of weather vanes and cupolas at discounts. Annapolis Weathervanes has a huge selection of styles and sizes. Free catalog. Phone: 800 724-2548. Fax: 410 757-8711. Website: www.weathervaneandcupola.com.
Weather Vanes, Cupolas

Antique Hardware & Home
19 Buckingham Plantation Dr., Bluffton, SC 29901
Replica hardware and accessories (many found nowhere else). 300 styles of door and cabinet hardware, weathervanes, tin ceilings, and cast iron barn bells, boot scrapers and horse hitches. Free Catalog. Phone: 800 422-9982, extension 1600. Fax: 803 837-9789. E-Mail: treasure@hargray.com. Website: www.antiquehardware.com.
Antique Hardware, New Hardware, Cupolas, Weather Vanes

Antique Woods & Colonial Restoration
1273 Reading Ave., Boyertown, PA 19512
Restoration and reproduction of colonial structures, timber frame barns, outbuildings, log homes and stone structures. Their service area includes PA, NY, NJ, CT, DE, MD, VA. They dismantle, re-erect and convert antique frames, and provide resawn siding and flooring. Free literature. Phone: 610 367-8193. Fax: 610 367-6911. E-mail: antiquewds@aol.com.
Vintage Timber Frames and Log Homes, Restoration Services, Custom Design

Allen Cupolas

Architectural Antiques Exchange
715 North 2nd Street, Philadelphia, PA 19123
Architectural salvage including doors, street lamps, leaded and beveled glass, signs, paneling and much more. Free catalog. Phone: 215 922-3669. Fax: 215 922-3680.
Architectural Salvage

Architectural Iron Company - Capital Cresting
PO Box 126, Milford, PA 18337
America's leading producer of roof cresting offers a complete line of lightweight, easy to install, economical to ship, unbreakable steel roof cresting, matching finials and snow guards. Custom sizes are available. Free catalog. Phone: 800 442-IRON. Fax: 717 296-IRON. Website: www.capitalcresting.com.
Finials, Cresting, Snow Guards

The Barn People

Architectural Reclamation, Inc.
312 South River Street, Franklin, OH 45005
Small family construction firm specializing in historic restoration and rehabilitation and serving southwest Ohio. They are experienced in structural repairs, log and timber frame work, masonry, custom wood-working, sheet-metal roofing and box gutters, plastering and more. Free literature. Phone: 513 746-8964. Fax: 513 746-7694. E-mail: mayapple@siscom.net.
Restoration Services

AshlandBarns
990-BH Butlercreek, Ashland, OR 97520
Blueprints are available for 98 barns, stables, garages with workshops, storage buildings, sheds and country homes - all designed for efficiency, beauty and economy of construction. Plans range in price from $12 to $55. Catalog: $5.00 (refunded with plan order). Phone: 541 488-1541. Website: www.ashlandbarns.com.
Building Plans (See the designs on pages 42 & 43)

The Barn People
PO Box 4, Morgan Hill, South Woodstock, VT 05071-0004
Offers a wide variety of antique Vermont barn frames which have been dismantled and restored and can be reassembled anywhere in the United States. The Barn People also offer custom-made, hand-hewn backyard office frames in kit form. Free literature. Phone: 802 457-3356. Fax: 802 457-3358. E-Mail: barnman@souer.net.
Vintage Timber Frames, Restoration Services, Custom Design, Building Kits

Barns by Country Woodshed
14800 Sweet Road, Peyton, CO 80831
Custom designed wood and metal barns built throughout southern Colorado. Gambrel barns are their specialty. They also provide help for "do-it-yourself" builders, by phone or at your site. Phone: 719 495-0510. Fax: 719 495-0510. E-mail: barns@pcisys.net.
Custom Barn Design and Building

Barns by Gardner, Ltd.
3833 West County Road 8, Berthoud, CO 80513
Custom design and building of pole barns and stables in northern Colorado. From simple loafing sheds to elegant showplace stables, Steve Gardner and his crew will build to the highest standards of quality, with the best materials, and at reasonable prices. Free literature. Phone: 970 532-3595. Website: www.barnsbygardner.com.
Custom Barn and Stable Design & Building, Building Plans

Barns by Gardner, Ltd.

BarnPlans
41-049 Ehukai Street, Waimanalo, HI 96795-1665
Simple, concise and easy-to-read blueprints designed with the owner/builder in mind. They have five sizes of Gambrel barn design: 16,' 20,' 24,' 28' and 32' widths with lengths designed to be modified to any required dimension. Phone or Fax: 808 259-7028. E-mail: dano@barnplans.com. Website: http://www.barnplans.com.
Building Plans for Homes, Garages, Shops and Barns (See the design on page 38)

Barn Works
PO Box 19, Peninsula Village Historic District, Peninsula, OH 44264
Barn Works is a collaborative of craftsmen dedicated to the preservation and rebirth of rural architecture. They specialize in new and antique timber frames and traditional joinery. They provide barn restoration and adaptive use, dismantling and reconstruction of historic structures, and historical research and documentation. Phone: 330 657-2135. E-mail: barnworks@juno.com.
Restoration Services, Vintage Timber Frames, New Timber Frames

Behm Design
23632 Highway 99, Suite F-112, Edmonds, WA 98026
Behm Design has prepared complete blueprints for a large variety of efficient homes, cottages, cabins, garages and utility buildings. Two catalogs are available: *Affordable Homes* and *Garage Plans*. The catalogs are $12.00 each. Phone: 800 210-6776. Website: www.behmdesign.com.
Building Plans (See the garage designs on pages 60 & 61)

Belcher's
2505 West Hillview Drive, Dalton, GA 30721
Pre-Civil War log cabins, weathered barn siding, split rails, hand-hewn beams. They consult on restoration of old log cabins. Free literature. Phone: 706 259-3482.
Architectural Salvage, Restoration Services

Donald J. Berg, AIA
P.O. Box 698, Rockville Centre, NY 11571-0698
Member of the American Institute of Architects and publisher of books on American country buildings, like this one. Custom design of new country buildings, historic research and renovations of old ones. Brochure of backyard barn blueprints, $4.00. Phone 516 766-5585. E-mail djberg@aol.com.
Custom Design, Historic Renovations, Building Plans (See the designs on pages 24, 28 & 35)

Berry Hill Ltd.
75 Burwell Road, St.Thomas, ON, Canada N5P 3R5
Country living and hobby farm specialists, offering garden equipment, antique reproduction lighting, hobby farm equipment and much more. Catalog: $3.00. Phone: 800 688-3072. Fax: 519 631-8935. E-mail: kfox@berryhill.on.ca. Website: www.berryhill.on.ca.
Cupolas, Light fixtures, Windmills, Weather Vanes

Bessler Stairway Company
3807 Lamar Ave., Memphis, TN 38118
A Bessler one-piece sliding stairway is an excellent choice for loft and attic access. Free catalog. Phone: 901 360-1900. Fax: 901 795-1253. E-Mail: bessler@bessler.com Website: www.bessler.com.
Pull-down Stairs

Better Barns
126 Main Street South, Bethlehem, CT 06751
Better Barns has been constructing upscale yard buildings in Connecticut for over twenty years. They pay meticulous attention to the details and quality of each of the small barns that they build and offer plans and hardware to homeowners outside of their area. Free catalog. Phone: 203 266-7989. Fax: 203 266-5352. E-mail: bbarns@wtco.net. Website: http://www.betterbarns.8m.com.
Construction of Small Barns and Sheds, Building Plans

BGS Plan Company
PO Box 1181, Roseville, CA 95678
Blueprints of farm and ranch barns, garages, stables, workshops and storage buildings are engineered to meet California and mountain state standards for seismic and wind resistance. The drawings are the most complete that you'll find, with accurate details, full-sized patterns and 3D framing isometrics that are keyed to complete lists of materials. Their $14.00 catalog presents over 1,200 building designs. Phone: 916 783-4332. Website: www.bgsplanco.com.
Barn, Garage, Stable and Workshop Building Plans (See the design on pages 62)

Big Spring Preservation Group, Inc.
1004 West Summer Street, Greenville, TN 37743
Specialists at dismantling and restoring antique log cabins, barns, outbuildings, garden sheds and timber frame barns and houses. The buildings can be reassembled anywhere. They offer a full line of antique building materials. Free product literature. Phone: 423 787-9373. Fax: 423 787-9312.
Vintage Timber Frames, Restoration Services, Garden Structures, Antique Cabins & Outbuildings

Board & Beam Co.
60 Wykeham Road, Washington, CT 06793
Barns and houses dismantled, rebuilt and restored. They sell salvaged materials including beams, planks, doors, hardware and architectural details. They also provide restoration and building shoring services. Free literature. Phone: 860 868-6789. Fax: 860 868-0721. E-mail: bbeams@ct1.nai.net.
Antique Timber Frames, Salvaged Lumber, Woodwork Details and Barn Doors, Antique Hardware, Restoration Services

Bob's Outhouses
246 Lower Cross Road, Nobleboro, ME 04555-8602
Traditional style outhouse kits. Besides their original purpose, these little buildings make great garden sheds, pool or pond changing rooms, and outdoor storage closets. Free literature. Phone: 800 654-8856.
Building Kits for Traditional Outhouses

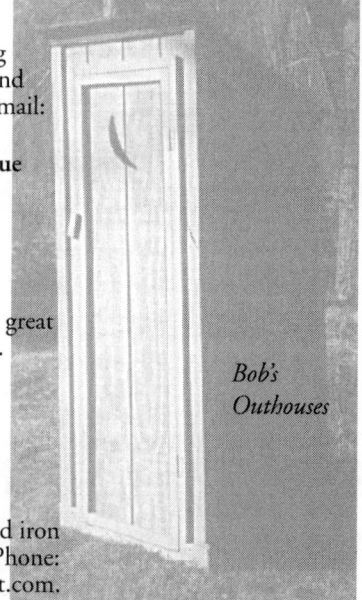

Bob's Outhouses

Bouvet USA, Inc.
540 De Haro Street, San Francisco, CA 94107
Decorative hardware company, established in 1884, offers a wide selection of hand-forged iron hardware. The products are made in France and adapted to US standards. Free catalog. Phone: 415 864-0273. Fax: 800-ATBOUVET. E-Mail: info@bouvet.com. Website: www.bouvet.com.
Antique and Custom Hardware

Bow House/Bow Bends
PO Box 900, Bolton, MA 01740
Makers of fine quality traditional cottages and garages plus exotic garden structures: bridges, gazebos, arbors, follies, privies and trellises. Catalog: $5.00. Phone: 978 779-6464. Fax: 978 779-2272.
Garden Structures, Building Kits

Brandywine Valley Forge
P.O. Box 1129, Valley Forge, PA 19482
Restoration blacksmiths specializing in hand-forged hardware, including strap hinges and pintles, hooks and hasps, barn door and gate bolts. They welcome custom work. Catalog; $5.00 or send a self-addressed stamped envelope for free literature. Phone: 610 948-5116. Fax: 610 948-5116.
Hand Forged Hardware

Brosamer's Bells
207 Irwin Street, Brooklyn, MI 49230
Bells for barn yards, backyards, cupolas, etc. Brass, cast iron and bronze, antique and new. Brosamer's is the country's largest dealer of pre-owned bells. Free product literature. Phone: 517 592-9030. Fax: 517 592-4511. Website: www.brosamersbells.com.
Barn and Yard Bells

CADSmith Studio
PO Box 526, Brookline, NH 03033
CADSmith Studio offers a selection of New England style designs for garages, barns, workshops and cottages at reasonable prices. Many of the barns have second floor apartments. All plans can be seen on their user-friendly website which has helpful links to other building sites. E-mail: bwsmith@cadsmith.com. Website: www.cadsmith.com.
Building Plans (See the designs on pages 54 & 55)

Cannonball: HNP
555 Lawton Ave., Beloit, WI 53512-0835
Cannonball: HNP is an 83 year old company that produces sliding door systems, tracks, trolleys, windows, walk doors, dutch doors, horse stalls, foil insulation, ventilators and cupolas for the agricultural marketplace. Free catalog. Phone: 800 766-2825. Fax: 800 834-7447. E-mail: cnbhnp@aol.com. Website: cnbhnp.com.
Doors, Windows, Hardware and Specialties for Barns, Stables and Agricultural Buildings

Cape Cod Cupola Co., Inc.
78 State Road, North Dartmouth, MA 02747
Cape Cod Cupola Company manufactures cupolas and weathervanes and specializes in custom work for both. They have a large selection of cupolas, weathervanes, sundials and house signs. Catalog: $2.00, refundable with first order. Phone: 508 994-2119. Fax: 508 997-2511.
Cupolas, Weather Vanes

Chestnut Oak Co.
3810 Old Mountain Road, West Suffield, CT 06093-2125
Chestnut Oak Co. erects new timber frame structures throughout New England and New York. They also restore historic buildings and dismantle, move, repair and erect old timber frame homes and barns. Product literature is free. Phone or Fax: 860 668-0382.
Vintage Timber Frames, New Timber Frames, Historic Preservation, Consultation

Christian & Son, Inc.
15022 Gearhart Road, Burbank, OH 44214
Specialists at designing, building and restoring barns, sheds and timber frame homes. Their new designs are based on the old ways of building and they work just as well today. Consultations are available. Phone: 330 624-7282. Fax: 330 624-0501. E-mail: rudad@aol.com.
Vintage Timber Frames, New Timber Frames, Restoration Services

CinderWhit & Company
733 Eleventh Avenue South, Wahpeton, ND 58075
Offers stock, replica or custom wood turnings, including porch posts, finials, newel posts, balusters and spindles for exterior and interior applications. Free brochure. Phone: 800 527-9064. Fax: 701 642-4204. Website: www.cinderwhit.com.
Wood Turnings

Cleary Building Corp.
PO Box 930220, Verona, WI 53593-0220
Cleary Building Corp. markets, designs and constructs pre-engineered laminated structures for agricultural, suburban or equestrian use. They serve the central US, the Midwest and the Northwest. Their wood frame structure is a unique design that's easily erected, affordable and flexible. Pre-painted exterior steel provides attractive, long lasting and low maintenance buildings. Each building is designed to the customer's specifications. Free literature. Phone: 800 373-5550. Fax: 608 845-7070. E-mail: sales@clearybuilding.com. Website: www.clearybuilding.com.
Pre-engineered Barns, Stables, Arenas, Garages and Agricultural Buildings

Colonial Barn Restoration
267 Old Bay Road, Bolton, MA 01740
Antique timber frame restoration and rebuilding, custom cupolas, cupola and steeple restoration and rebuilds, structural member replacement using mortise and tenon joinery. They serve the northeast USA. Phone or Fax: 978 779-9865. E-mail: tmurphy@aol.com.
Restoration Services, Cupolas

Colonial Cupolas, Inc.
1816 Nemoke Trail, P.O. Box 38, Haslett, MI 48840
America's largest selection of cupolas, assembled or as kits. Catalog: $3.00. Product Literature: $3.00. Phone: 517 349-4408.
Cupolas, Weather Vanes, Sundials, Cast Metal Date and Street Number Plaques

Connolly & Co. Timber Frame Homes and Barns
10 Atlantic Highway, Edgecomb, ME 04556
Connolly & Co. restores existing timber frame structures; custom designs, cuts and erects homes, barns, outbuildings, additions, and truss systems; and offers four styles of pre-cut barn kits. Free literature. Phone: 207 882-4224. Fax: 207 882-4247. E-mail: connolly@lincoln.midcoast.com. Website: www.connollytimberframes.com.
Vintage Timber frames, New Timber Frames, Restoration Services, Barn Building Kits

Coppercraft, Inc.
2143 Joe Field Road, Suite 100, Dallas, TX 75229
Coppercraft utilizes traditional metalworking skills and modern technology to create high quality architectural sheet metal products including cupolas, spires, weathervanes and more. Free catalog and product literature. Phone: 800 486-2723. Fax: 972 484-3008. E-Mail: info@coppercraft.com Website: www.coppercraft.com.
Cupolas, Finals, Spires, Gutters, Vents, Roofing

The Copper House
1747 Dover Road (Route 4), Epsom, NH 03234-4416
Offers brass and copper interior and exterior lighting and copper weather vanes. All products are made in New Hampshire. Lighting is U.L. approved. Catalog: $4.00. Phone: 800 281-9798. Fax: 603 736-9798. Website: www.thecopperhouse.com.
Post Lamps, Carriage House Lamps, Weather Vanes

Country Carpenters, Inc.
326 Gilead Street, Hebron, CT 06248
Designers and manufacturers of fine pre-cut, New England style post and beam barns, carriage houses, garages and sheds. Catalog: $4.00. Phone: 860 228-2276. Fax: 860 228-5106. Website: www.countrycarpenters.com.
Building Kits (See the designs on pages 31 & 40)

Country Design
PO Box 774, Essex, CT 06426
Designers of traditional New England style buildings. Blueprints include cottages, carriage houses, barns, garages, garage apartments, stables, sheds, pool houses, well houses, gazebos and period fence designs. Their complete catalog is $8.00, by mail.
Building Plans (See the designs on pages 32, 33, 34, 41 & 56)

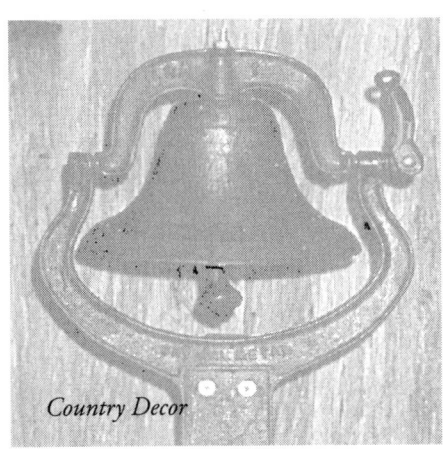

Country Decor

The Copper House

Country Decor
8015 Green Devon Drive, Houston, TX 77095
Country Decor sells old fashioned farm and backyard bells. The bells are made in the U.S.A., and are newly cast with antique looks and loud rings. Other products include weather vanes, boot scrapers and horse hitches. Free product literature. Phone: 800 888-2196. Fax: 281 550-8658. E-mail: farmbells@aol.com. Website: www.countrydecor.com.
Farm and Yard Bells, Weather Vanes, Horse Hitches

Country Settings, Inc.
3305 West 4th Ave., Suite C, Belle WV 25015
Specialists in the recovery of authentic 150 year old hand-hewn log cabins and barns. They have a large inventory of antique log cabins and barns. Any of them can be reassembled on your site. Other vintage materials include resawn chestnut and oak, hand-cut stone, split-rail fence and weathered barnboard. Phone: 304 925-3863. Fax: 304 925-3303. E-mail: handhewn@aol.com. Website: www.countrysettings.com.
Vintage Hand Hewn Log Cabins, Antique Timber Frames and Building Materials, Restoration Services

Craftwright Incorporated
100 Railroad Ave., Suite 105, Westminister, MD 21157
Custom, hand-crafted timber frames. Antique timbers and frames are available. Phone: 410 876-0999.
Vintage Timber Frames, New Timber Frames

Crosswinds Gallery, Inc.
29 Buttonwood Street, Bristol, RI 02809
Offers a large selection of quality weather vanes, cupolas and finials in a variety of materials and at a variety of prices. Crosswinds Gallery specializes in custom design and crafting. Imagine a weather vane, and they'll make it for you. Their extensive catalog of designs is free. Phone: 401 253-0334. Fax: 401 253-2830. E-mail: wvanes@aol.com. Website: www.crosswinds-gallery.com.
Weather Vanes, Finials, Cupolas, Custom Design

Cumberland General Store
#1 Highway 68, Crossville, TN 38555
General merchandise catalog with old-time and country specialties and restoration and building products. Weathervanes, boot scrapers, hitching posts, farm bells, farmstead tools, hardware, country home plans, pumps, windmills and building books. Catalog: $4.00. Phone: 931 484-8481. Fax: 931 456-1211. E-mail: generalstore@worldnet.att.net Website: www.cumberlandgeneral.com.
Hardware, Books, Building Plans, Weather Vanes

Cumberland Woodcraft Company
P.O. Drawer 609, 10 Stover Drive, Carlisle, PA 17013
Provides interior and exterior Victorian millwork, gables, balustrades, corbels and brackets, mouldings and custom items. Catalog: $5.00. Phone: 800 367-1884. Fax: 717 243-6502. E-Mail: cwc@pa.net Website: www.cumberlandwoodcraft.com.
Woodwork, Finials & Cresting

Dad's Woodshop
19392 Renwood Ave., Euclid, OH 44119
Dad's is a full service custom woodshop focusing on specialty and hard-to-find products. Product literature: $1.00. Phone: 216 383-8808.
Woodwork, Cupolas, Arbors, Trellises

Directory 73

Dalton Pavilions, Inc.
20 Commerce Drive, Telford, PA 18969
Dalton Pavilions Inc. offers fine western red cedar prefabricated pavilions and garden structures that can be shipped throughout the U.S. and internationally. Free catalog. Phone: 215 721-1492. Fax: 215 721-1501.
Prefabricated Garden Structures

Davis Frame Company
PO Box 1079, Claremont, NH 03743
Specialists at designing and hand-crafting timber frames for homes, barns and additions in new Douglas fir, pine, oak or reclaimed timbers with mortise and tenon joinery, chamfered edges and curved braces. Packages include the frame, stress skin panels, exterior finish materials, windows and doors. Free flyer. Catalog and Plan Portfolio: $15.00. Phone or Fax: 800 636-0993. E-mail: inquiry@davisframe.com. Website: www.davisframe.com.
New Timber Frame Building Kits, Custom Design

Denninger Weather Vanes & Finials

Denninger Weather Vanes & Finials
77 B Whipple Road, Middletown, NY 10940
Weather vanes featuring finely hand-crafted horses, roosters, eagles, banners, scrolls, arrows, caps and finials. They offer custom and standard designs, farm and business logos, and historic replications. They have an informative website for anyone interested in the art and lore of weather vanes. Free literature. Phone or Fax: 914 343-2229. E-mail: al@denninger.com. Website: www.denninger.com.
Weather Vanes, Finials

Designer Doors, Inc.
283 Troy Street, River Falls, WI 54022
Designer Doors are garage doors handcrafted to match architectural elements of distinctive homes. Many appear to swing, fold or slide open, yet all conveniently roll up with automatic openers. Phone: 800 241-0525. Fax: 715 426-49999. E-mail: info@designerdoors.com. Website: www.designerdoors.com.
Carriage House and Garage Doors

Gardensheds

Dreaming Creek Timber Frame Homes, Inc.
2487 Judes Ferry Road, Powhatan, VA 23139
Custom timber framing and design of homes, churches, covered bridges, gazebos and barns. They erect frames, install structural panels, and supply timber, oak plank flooring and paneling. Free literature. Literature and video: $10.00. Phone: 804 598-4328. Fax: 804 598-3748. E-mail: dctfh@aol.com. Website: www.dreamingcreek.com.
New Timber Frames, Timber Frame Trusses, Custom Design, Garden Structures

Eagle Creek Designs, Inc.
6025 Schustrich Road, P.O. Box 163, Mantua, OH 44255
Eagle Creek Designs stocks log houses, timber frames, cut sandstone, ornamental stone, mantles, flooring, hardware, beadboard and beams. Phone 330 274-2041 for more information.
Building Materials, Vintage Timber Frames and Log Homes

Emerald Woodworking
21 Elbormar Drive, Huntington, NY 11743
Offers cedar ventilating louvers in any shape or size, exterior shutters, rough-hewn pine planking, board and batten doors and custom woodwork. Their service area is New York and New England. Free literature - send S.A.S.E. to the address above. Phone: 516 754-0377. Fax: 516 935-7546.
Woodwork

William T. Hardy, Builder

Eugenia's Antique Hardware
5370 Peachtree Road, Chamblee, GA 30341
Provides authentic antique hardware, including door and furniture hardware, hinges, handles, latches, knockers, mechanical bells and forged iron strap hinges. Catalog: $1.00. Phone: 800 337-1677. Fax: 770 458-5966. Website: eugeniaantiquehardware.com.
Antique Hardware

Fingerlakes Weathervanes and Cupolas
P.O. Box 554, Canandaigua, NY 14424
Unique copper and brass weathervanes, made in the U.S.A. They create combination brass and copper American flag weather vanes. Free catalog and product literature. Phone: 716 394-1091.
Weather Vanes, Finials & Cresting

Garden Oaks Specialties
1921 Route 22 West, Bound Brook, NJ 08805
Manufacturers of high quality, prefab garden buildings including sheds, gazebos, arbors, playhouses and mailboxes in a variety of styles. Buildings are shipped complete to your backyard. Free literature. Phone: 800 590-7433. Fax: 732 356-7202. E-mail: info@gardenoaks.com Website: www.gardenoaks.com.
Prefabricated Sheds and Garden Buildings

Gardensheds
651 Millcross Road, Lancaster, PA 17601
Classic designs of outbuildings and garden sheds. Complete buildings, hand-crafted of fine woods, are delivered to your property. Custom designs and building plans are also available. Portfolio: $5.00. Phone: 717 397-5430. Fax: 717 397-0217. Website: www.gardensheds.com.
Prefabricated Sheds and Garden Buildings, Custom Design, Building Plans

Great Northern Barns
Box 912E, RFD 2, Canaan, NH 03741
Great Northern Barns works with all aspects of timber framing with an emphasis on providing and erecting antique barn frames. Free literature. Video: $10.00. Phone: 603 523-7134. Fax: 603 523-7134. E-mail: ejl@endor.com. Website: www.greatnorthernbarns.com.
Vintage Timber Frames

Gothic Arch Greenhouses
P.O. Box 1564-BB, Mobile, AL 36633
Gothic Arch Greenhouses, a division of Trans-Sphere Trading Corp., creates beautiful Gothic style greenhouse kits in redwood, with polycarbonate glazing, in sizes for hobby, backyard or commercial use. Designs are freestanding or lean-to. They also provide heating, cooling, ventilation and humidification systems for greenhouses. Product literature: $2.00. Phone: 334 432-7529. Fax: 334 432-7972. Website: www.zebra.net/~gothic/.
Greenhouse Kits and Environmental Systems

Green Star Forge
3 Myrtle Street, Taunton, MA 02780-4111
One-man shop specializing in custom forged iron work. Catalog: $2.00. Phone: 508 824-3077.
Hand Forged Iron Hardware

Habitat Post & Beam
21 Elm Street, South Deerfield, MA 01373
Habitat Post & Beam pre-manufacturers material packages for post and beam homes, barns and additions. Their in-house design and engineering services allow you to custom design your project to fit your personal vision. They can work from your ideas, your architect's design or a plan from an extensive library of designs. Catalog: $12.00 (it's downloadable at their website). Phone: 800 992-0121. Fax: 413 665-4008. E-mail: sales@postandbeam.com. Website: www.postandbeam.com.
Building Kits, Manufactured Buildings, New Timber Frames, Custom Design

Hager Companies
139 Victor Street, St.Louis, MO 63104
Manufacturer of hinges and builders' hardware including barn door rollers, tracks and accessories, strap hinges, shed hardware, metal thresholds, door sweeps and gate hardware. Free catalog and product literature. Phone: 314 772-4400. Fax: 314 772-0744. Website: www.hagerhinge.com.
Hardware, Barn Door Rollers & Track

Hahn Woodworking Company, Inc.
109 Aldene Road, Roselle, NJ 07203
Custom wooden garage doors built to your specifications. They offer stile and rail doors with flat or raised panels, historic barn and carriage-house style doors with convenient overhead motorized operation, traditional swing-out doors and sliding doors. Free catalog and product literature. Phone: 908 241-8825. Fax: 908 241-9293.
Barn & Garage Doors, Custom Design

Hahn Woodworking Company, Inc.

J.W. Hall Enterprises, Inc.
PO Box 68, Santa Fe, TX 77517
Designers and manufacturers of horse stalls, exercise arenas, barns, and accessories since 1982. All are designed for safety and convenience and built to last a lifetime from heavy gauge hotdipped galvanized steel. Free catalog and product literature. Phone: 800 475-8158. Fax: 409 925-4782. E-mail: jimhall@sat.net. Website: www.jwhall.com.
Custom Equestrian Structures, Stalls, Stable Equipment

Handy Home Products
6400 E.11 Mile Road, Warren, MI 48091
North America's largest manufacturer of ready-to-assemble wooden storage buildings, gazebos, timber buildings, solar sheds and playhouses. Their buildings are sold through home centers. Free catalog. Phone: 800 221-1849. Fax: 810 757-6066. Website: www.handyhome.com.
Building Kits for Cabins, Garages, Storage Sheds, Solar Sheds, Gazebos and Playhouses

J.W. Hall Enterprises, Inc.

William T. Hardy, Builder
Rural Route 2, Box 344, North Bennington VT 05257
Custom builder of country homes and barns. Hardy provides fine craftsmanship in pole-framing, light-frame and log construction in southern Vermont, northwestern Massachusetts and eastern New York. Phone: 802 442-4075
Custom Builder

Historic Housefitters Co.
P.O. Box 26, 32 Centre, Route 312, Brewster, NY 10509
Offers hand-forged iron hardware of all types, including strap hinges and pintles, cane bolts, thumb latches, door pulls. Stock hardware, custom designs and reproductions. Catalog: $3.00. Phone: 914 278-2427. Fax: 914 278-7726. Website: historichousefitters.com.
New Hardware, Custom Reproductions

Homestead Design, Inc.
P.O. Box 2010, Port Townsend, WA 98368
Homestead Design's buildings feature flexible plans and simple framing. Mail-order plans are easy to understand, easy to use, and include a list of the materials you'll need to build the structure and exterior shell. Interior finishes and a variety of interior layouts are your choice. You can build these designs just the way you want. Catalog: $5.00. Phone or Fax: 360 385-9983. Website: www.homesteaddesign.com.
Building Plans (See the designs on pages 36, 37, 44, 58, & 59)

Handy Home Products

H.T. Cadd & Blueprint Service
N 8939 Townline Road, East Troy, WI 53120
Complete blueprints for bridges, cupolas, privies, gazebos, garden and backyard structures. Free illustrated brochure of books and catalogs. phone: 800 996-3002. Fax: 414 642-3002. E-mail: htcadd@netwurx.net. Website: www.htcadd.com.
Building Plans for Cupolas and Garden Structures (See the cupola designs on page 66)

Ken Hume, Engineer
Oakhurst, Sherfield Road, Bramley, Hampshire, England RG26 5AQ
Registered professional engineer practicing in the USA and UK. Timber-frame designer and structural analyst with an understanding of traditional practices in America and Europe. Phone: 011 44 1256 881344. E-mail: ken.hume@pareuro.com.
Timber Frame Engineering, Restoration Services

Independent Protection Co., Inc.
1603-09 South Main Street, P.O. Box 537, Goshen, IN 46527
Ornamental and conventional lightning protection equipment, systems and products for all types of homes, barns and related structures. Catalog: $10.00. Product literature: $5.00. Phone: 219 533-4116. Fax: 219 534-3719. E-Mail: ipc@netbahn.net.
Lightning Rods, Weather Vanes

Independent Protection Company

Iron Intentions Forge
RD#2, Box 2399C, Spring Grove, PA 17362
Custom forged hardware and accents in steel, stainless, brass, copper and aluminum. Phone: 717 229-2665.
Antique Hardware, New Hardware, Weather Vanes

Ives Weathervanes
Box 101A, RR1, Charlemont, MA 01339
Hand-formed, elegant custom copper and brass weather vanes. Ives Weathervanes creates three-dimensional hammer formed vanes with "chased" in details and two-dimensional silhouette style pieces. Gold leafing is available. Catalog: $1.00. Phone: 413 339-8534.
Weather Vanes

Jack's Country Store
P.O. Box 710, Bay Avenue & Highway 103, Ocean Park, WA 98640
Jack's genuine Alladin kerosene lamps are smokeless, odorless and as bright as a 60-watt light bulb. The lamps are great for nonelectric buildings. Jack's offers a complete selection of lamps and parts. Catalog: $1.00. Phone: 360 665-4988. Fax: 360 665-4989.
Kerosene Lamps, Hardware

Jaderloon
PO Box 685, Irmo, SC 29063
or 1320 South Burleson Blvd., Burleson, TX 76028
Hobby and commercial greenhouses and greenhouse supplies, carts and wagons. Sizes range from an acre plus down to 8'x 8'. Free Home Greenhouse Planning Guide. Commercial greenhouse catalog - $3.00. Phone: 800 258-7171. Fax: 803 798-6584. E-mail: jaderloon@aol.com. Website: www.jaderloon.com.
Greenhouses

Larry James Designs
2208 Justice Street, Monroe, LA 71201
Home designs that reflect Louisiana and Southern traditions. A catalog of 64 home and cottage blueprints is available for $10.00. A website presents even more designs. Phone: 800 742-6672. Website: www.larryjames.com.
Building Plans (See the design on page 47)

Just Outbuildings
P.O. Box 42, Brewster, NY 10509
Just Outbuildings produces complete architectural plans for garages, sheds, pool houses and garden buildings in a variety of sizes. Styles range from contemporary to traditional. Plans can be customized to suit your needs. Catalog: $6.00. Phone: 914 279-4542. E-Mail: gjgaspar@bestweb.net
Building Plans, Custom Design (See the designs on pages 64 & 65)

Kalglo Electronics Co., Inc.
5911 Colony Drive, Bethlehem, PA 18017-9348
Kalglo's electric infrared heaters provide safe, efficient spot heating of people and animals in barns, garages and workshops. Free product literature. Phone: 888 452-5456. Fax: 610 837-7978. E-mail: kalglo@kalglo.com. Website: www.kalglo.com.
Electric Infrared Heaters

Ives Weathervanes

Kayne & Son Custom Hardware, Inc.
100 Daniel Ridge Road, Chandler, NC 28715
Forges custom iron hardware including strap hinges, latches, bolts, braces, door rollers, hasps, locks, and branding irons. Provides cast brass and bronze latches, pulls and fasteners in various finishes. Kayne & Son are experts in restoration, repair and reproduction of antique hardware. Catalog: $5.00. Phone: 704 667-8868. Fax: 704 665-8303. E-Mail: kaynehdwe@ioa.com.
New Forged Hardware, Restoration Services

Kool-O-Matic Corporation
1831 Terminal Road, Niles, MI 49120
Manufacturers of automatic power attic and space ventilators. Their cupola-matic power ventilators are offered with automatic controls, thermostat and humidistat and are available with white or natural redwood cupolas. Free literature. Phone: 616 683-2600. Fax: 616 683-2318. E-mail: Fanskom@aol.com.
Ventilators, Cupolas

Kayne & Son Custom Hardware, Inc.

Landmark Services, Inc.
7 Oakland Street, Medway, MA 02053
Landmark Services, Inc. is a restoration and renovation general contracting firm specializing in the restoration of historic homes, barns and churches.sembled and reassembled. Free literature. Phone or Fax: 508 533-8393. E-Mail: landmark@gis.net. Website: www.landmarkservices.com
Restoration Services

Lehman's
One Lehman Circle, P.O. Box 41, Kidron, OH 44636
Lehman's 160 page nonelectric catalog contains 2,500 items you thought weren't made any more, including farm tools, wood stoves, grain mills, butter churns, copper kettles and more. Catalog: $3.00. Phone: 330 857-5757. Fax: 330 857-5785. E-Mail: info@lehmans.com. Website: www.lehmans.com.
Hardware, How-to Books, Tools, Gas Lamps

Lemee's Fireplace Equipment
815 Bedford Street, Bridgewater, MA 02324
Provides wrought-iron hardware, strap hinges, boot scrapers, barn bells and gongs, hitching posts and weather vanes. Catalog: $2.00. Phone: 508 697-2672.
Antique Hardware, New Hardware, Weather Vanes

Kool-O-Matic Corporation

Lester Building Systems

Lester Building Systems
1111 2nd Avenue South, Lester Prairie, MN 55354
Lester Building Systems manufactures pre-engineered wood-frame structures for agricultural, equestrian, commercial and backyard use. Buildings are sold and erected by 450 independent, locally-owned dealers from the Rockies to the East coast. Free product literature is available. Phone: 800 826-4439. Fax: 320 395-5395. E-Mail: info@lesterbuildingsystems.com. Website: www.lesterbuildingsystems.com.
Pre-engineered Buildings, Custom Design

Louisiana Country Homes
c/o Poole Lumber, PO Drawer 1240, Covington, LA 70734
Houses, cottages and barns designed in Louisiana Acadian, Creole and vernacular styles by Bob Sander, AIBD. Complete blueprints are available for all. Illustrated literature is free. Phone: 800 525-0006.
Building Plans

The Mailbox Shoppe
2566A Hempstead Turnpike, East Meadow, NY 11554
The Mailbox Shoppe represents over 30 manufacturers of weathervanes, cupolas, mailboxes, mailbox posts, custom cast signs and other home accessories. Free catalog and product literature. Phone: 800 330-3309. Fax: 516 735-6191. E-mail: sales@mailboxnet.com Website: www.mailboxnet.com.
Mailboxes, Cupolas, Weather Vanes

The Mailbox Source
12367 Deerbrook Lane, Los Angeles, CA 90049-1909
Residential freestanding and wall-mounted mailboxes in a wide variety of styles and materials. They offer locking boxes, large capacity boxes and novelty boxes. Free catalog. Phone or Fax: 800 209-0111.
Mailboxes

Martha's Vineyard Plans
PO Box 350, Vineyard Haven, MA 02568
Blueprints of cottages, homes, small barns, garages, sheds, guest houses, gazebos, decks and more. Contempory and traditional designs all reflect Martha's Vineyard and New England architectural heritage. Catalog: $15.95 plus $3.00 postage. Phone: 888 847-5267. Website: www.vineyard.net/biz/mvplans.
Barn, Home and Outbuilding Plans, Custom Design (See the designs on pages 26 & 46)

Mink Hill Timber Frame Homes, Inc.

Sam Marts Architects and Planners/ White Oak Timber Frame
2104 West Wabansia, Chicago, IL 60647
Professional architects and builders provide integrated design and timber frame construction services for clients, worldwide. Phone: 773 862-0123. Fax: 773 862-0173. E-mail: info@timbersmart.com. Website: www.timbersmart.com.
Custom Design, New Timber Frames

Meyers Restoration & Architectural Salvage
R2 Box 1250, East River Road, Clinton, ME 04927
Restoration, preservation and reproduction of 18th and 19th century buildings. They provide carpentry services from frame repair to reproduction millwork, period consulting and planning. They also buy and sell antique architectural details and building materials. Free literature. Phone: 207 453-7010. Fax: 207 238-9905. E-mail: myrest@mint.net.
Architectural Antiques, Antique Timber Frames, Woodwork, Restoration Services

John T. Miller, Barnbuilders
229 Church Street, East Harwich, MA 02645
New England style barns, garages, sheds, pool houses, small bridges and related breezeways and connectors built on Cape Cod, South East Massachusetts and New Hampshire. Miller Barnbuilders work in modified timber frame construction. They try to incorporate interesting salvaged millwork and hardware into their projects to give each a one-of-a-kind look. Project photos available on request. Phone: 508 430-0684.
Custom Barn Builders, Restoration Services

John T. Miller, Barnbuilders

The Millworks, Inc.
P.O. Box 2987, Durango, CO 81302
The Millworks Inc. offers Victorian, traditional, country and southwest millwork. Catalog: $2.00. Phone: 970 259-5915. Fax: 970 259-5919.
Woodwork

Mink Hill Timber Frame Homes, Inc.
285 Davis Road, Bradford, NH 03221
Builder of timber frame homes and barns. Mink Hill provides architectural design and engineering services, supplies antique barns and restores existing barns. Free literature. Phone or Fax: 603 938-5203. E-mail: kwhitehead@conknet.com.
New Timber Frames, Vintage Timber Frames, Custom Design, Restoration Services

Mountain Construction Enterprises
PO Box 1177, Boone, NC 28607
Builds timber frame, log, and other custom buildings. They will ship building kits nationwide or build complete homes in North Carolina. Free literature. Catalog: $12.00. Phone: 828 264-1231. Fax: 828 264-4863. E-mail: mtnconst@boone.net. Website: www.mountainconstruction.com.
Log and Timber Frame Building Kits, Custom Design

National Horse Stalls
PO Box 153, Raphine, VA 24472
National Horse Stalls provides horse care and barn equipment. Products include stall components, dutch doors, aisle doors, stall flooring, horse stocks, feeders, waterers, window grills, exercisers and arena footing material. Free product literature. Phone: 800 903-8908. Fax: 540 337-5973. E-mail: horsstal@cfw.com. Website: www.nationalhorsestalls.com.
Stable Equipment

National Horse Stalls

New Energy Works Timberframers
1755 Pioneer Road, Shortsville, NY 14548
New Energy Works Timberframers designs and builds timber frame homes, barns and other structures of the highest quality. Their projects represent a philosophy of individual attention to the client and to the building process. Free literature. Phone: 716 298-3220. E-mail: jonathan@newenergyworks.com. Website: www.newenergyworks.com.
New Timber Frames, Custom Design, Vintage Timber Frames.

New England Barn Company
63 Gaylord Road, Gaylordsville, CT 06755
Creates precut timber frame barns fashioned in classic New England styles, with authentic mortise and tenon joinery. A variety of sizes and styles are available, and custom work is welcomed. Free literature. Phone or Fax: 860 350-5544.
Barn Kits, New Timber Frames

New England Barn Company

New England Cupola
184 Mattapoisett Road, Acushnet, MA 02743
Builder of fine, hand-crafted cupolas in a wide range of styles and sizes. Custom work is New England Cupola's specialty. Free catalog and product literature. Phone: 508 995-5331. Fax: 508 998-7041.
Cupolas, Weather Vanes, Custom Design

New England Outbuildings
P.O. Box 621, Westbrook, CT 06498
New England Outbuildings creates new farm and garden outbuildings in the traditional manner by meticulously preserving the lines, proportion and details of New England's historic buildings. Post and beam frames are milled from oak, cut with mortise and tenon joints and shipped to your site to be assembled with wooden pegs. Frame kits include designs for barns, wagon sheds, corn cribs and more. Free literature. Phone: 860 669-1776.
New Timber Frames, Building Kits

New Old Products

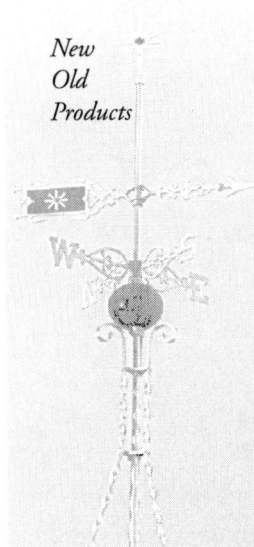

New Jersey Barn Company
PO Box 702, Princeton, NJ 08542
New Jersey Barn Company offers antique oak barn timber frames which they re-erect on your property. Free literature. Phone: 609 924-8480. Fax 609 730-1030.
Vintage Timber Frames

New Old Products, Inc.
PO Box 1272, Kokomo, IN 46903
New Old Products Inc. provides antique style lightning protection. They provide parts for repair and restoration of old systems and complete new systems with modern protection and performance. The systems can be U.L. approved. Parts include historic reproduction points, glass balls, arrows, weather vanes and compass points. Free product literature. Phone or Fax: 765 868-3092. E-mail: newoldproducts@townwire.com. Website: www.newoldproducts.com.
Antique Style Lightning Protection Systems & Restoration

Niagara Designs, Inc.
PO Box 191, Niagara Falls, ON, Canada L2E 6T3
Niagara Design, Inc. provides stock construction plans for garden structures, including gazebos, sheds, playhouses, greenhouses, garages and more. Plans can be used for do-it-yourself projects or by your builder. E-mail: plans@niagaradesigns.com. Website: www.niagaradesigns.com.
Building Plans

Niff-Tone Distributors, Inc.
10274 West 600 South, Mentone, IN 46539
Manufacturers of Air Flow Ridge Ventilators which feature corrosion resistant fiberglass construction and simple gravity operation. These durable vents are installed in conjunction with normal ridge material. They provide effective, low cost air flow. Free product literature. Phone: 800 458-0840. Fax: 219 353-7183.
Ventilators

North Woods Joinery
PO Box 1166, Burlington, VT 05402-1166
Creates traditional post and beam structures, including gazebos, barns, sheds and homes. Your choice of wood species includes pine, hemlock, oak and Douglas fir. Free product literature. Phone: 802 644-2400 or 802 644-2500. Fax: 802 644-2509.
New Timber Frames, Building Kits

O'Brock Windmills
9435 12th Street, North Benton, OH 44449
O'Brock Windmills sells and installs old style water pumping windmills which are very often found next to old country homes, carriage houses and barns. They were, and still are, used to provide water for livestock. Today, many people install them just to look good turning in the breeze. Catalog: $2.00. Phone: 330 584-4681. Fax: 330 584-4682. E-Mail: windmill@cannet.com.
Water Pumping Windmills, Hand Pumps, Hydraulic Rams

David D. Parker, Structural Restoration
904 Upper Dummerston Road, Brattleboro, VT 05301
Provides restoration consulting and contract services, 18th and 19th century timber frames and antique lumber. Phone: 802 257-5717. Fax: 802 257-5719. E-mail: sjmpr@sover.net. Website: parkerrestoration.com.
Restoration Services, Vintage Timber Frames, Antique Building Materials

Rapid River Rustic, Inc.
PO Box 10, Rapid River, MI 49878
Rapid River Rustic manufactures cedar log homes. They specialize in custom design and can provide a full set of blueprints and material to build your cedar log home. Free literature. Catalog: $10.00. Phone: 800 422-3327. Fax: 906 474-6500. E-mail: rrrustic@up.net. Website: www.rapidriverrustic.com.
Log Home Building Kits, Custom Design

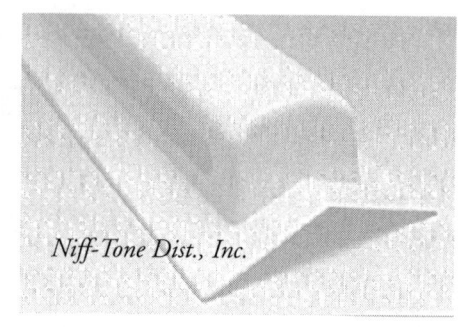
Niff-Tone Dist., Inc.

Renovator's Supply
PO Box 2515, Conway, NH 03818-2515
Looking for something special? Renovator's Supply sells hard-to-find items for building and restoration projects. Products include wrought iron hardware, cupolas, weather vanes, mail boxes, garden furniture, light fixtures, reproduction materials, hardware and plumbing fixtures. Free catalog. Phone: 800 659-2211.
Reproduction Hardware and Building Products

Restoration Resources
167 Dock Road, Alna, ME 04535
Specializes in period restoration of 18th century buildings in the state of Maine. Professional services include consultation, house inspection, fine carpentry, and restoration or relocation of early structures. Free literature. Phone: 207 586-5680. E-mail: fossel@oldhouserestoration.com. Website: www.oldhouserestoration.com.
Restoration Services, Vintage Timber Frames, Antique Hardware and Building Components

Recycled Products Company
18294 Amber Road X44, Monticello, IA 52310-7708
Manufacturer of plastic lumber and "100 Year" windows, recycled from milk containers. White, venting barn windows are USDA approved, never need paint or putty, come in a variety of efficient, attractive sizes and help reduce landfill waste. Free product literature. Phone: 800 765-1489. Fax: 319 465-1489.
Barn Sash and Windows, Recycled Lumber

Rockin J Horse Stalls
PO Box 869, Mannford, OK 74044
Designers and builders of fine horse stalls for over 18 years. Custom designs to suit your needs. Free catalog. Phone: 800 765-7229. Fax: 918 865-4191. E-mail: info@rockinjhorsestalls.com. Website: www.rockinjhorsestalls.com.
Custom Horse Stalls

David D. Parker, Structural Restoration

McKie Roth Design, Inc.
P.O. Box 31, Castine, ME 04421
You'll find some of McKie Wing Roth's barn designs in this book, but he is best known for his traditional New England home designs. His folio of homes features twenty-eight Capes, Colonials, Saltboxes and Gambrels, all with contemporary interior layouts for modern living, and all with construction plans you can order. Catalog: $18.00. Phone: 800 232-7684. Fax: 207 326-9513. Website: www.mckieroth.com.
Building Plans, Traditional Home and Barn Designs (See the designs on pages 50, 51 & 63)

Royalston Oak Timber Frame

Royalston Oak Timber Frame
122 North Fitzwilliam Road, Royalston, MA 01331
New England and medieval English timber frames in oak, pine, hemlock and fir. The artisans of Royalston Oak work with the highest quality timbers and provide traditional joinery. Catalog: $8.00. Free literature. Phone: 800 317-1129. Fax: 978 249-9633. E-mail: tmusco@hotmail.com.
New Timber Frames, Medieval English-Style Timber Frames

Salter Industries
P.O. Box 183, Eagleville, PA 19408
Manufacturer of steel and wood spiral stairs in a variety of sizes and designs including half-turns and units that meet BOCA and UBC code requirements. Free literature. Phone: 610 631-1306. Fax: 610 631-9384.
Spiral Stairs

Second Harvest Salvage
RR#1, Box 194-E, Jeffersonville, VT 05464
Provides antique hand-hewn barn and house frames, wide board flooring, beams and other antique building materials. Second Harvest also provides consultation on building restoration. Call for information. Phone: 802 644-8169.
Vintage Timber Frames, Restoration Services

David Shea

David Shea
48 Center Street - Route 322, Wolcott, CT 06716
Dave Shea has an inventory of approximately 50 antique horse-drawn vehicles for sale for commercial or pleasure use, or for collectors. Wagons, carriages, carts and sleds range from draft horse to pony sizes. Antique stable appointments are also available. For information, phone 203 879-3169
Antique Horse Drawn Vehicles, Stable Equipment

Sheldon Designs, Inc.
1330 Route 206 - #204, Skillman, NJ 08558
Architect Andy Sheldon has created a line of barns, sheds, garages and country cabins and cottages from one-room wilderness getaways to elegant shingle-style vacation homes. Detailed building plans are available for all. Some can also be purchased as easy-to-build log kits. Catalog: $7.00. Phone: 800 572-5934. Fax: 609 683-5976. E-mail: andysheldon@worldnet.att.net. Website: sheldondesigns.com.
Building Plans, Custom Design, Building Kits (See the designs on pages 27, 48 & 49)

Shelter-Kit Incorporated
22 Mill Street, Tilton, NH 03276
Pre-cut post and beam houses, cabins and multi-purpose barns are sold in kit form. These buildings are designed specifically for owner assembly. No construction skills or power tools are needed. Free literature. Phone: 603 286-7611. Fax: 603 286-2839. E-mail: buildings@shelter-kit.com. Website: www.shelter-kit.com.
Country Home, Cabin and Barn Building Kits (See the design on page 52)

Singletree & Associates
19840 Rocking Horse Road, Bend, OR 97702
Restoration and repair of log, timber frame and conventionally framed structures. Services are provided from assessments through "hands-on" construction. Singletree & Associates also builds new barns and outbuildings. Phone: 541 382-7143.
Log and Timber Frame Building, Restoration Services

Singletree & Associates

Spiral Stairs of America, Inc.
1700 Spiral Court, Erie, PA 16510-1367
Manufacturer of spiral, curved and straight stair systems for indoor and outdoor use. Stairs are made of steel, wood or aluminum. Free product literature. Phone: 800 422-3700. Fax: 814-899-9139.
Stairs

Strictly Barns, Inc.
2017 Route 78, Java Center, NY 14082
Strictly Barns, Inc. is your one stop barn shop. They offer new custom designed Stockade pre-engineered pole structures, doors, windows, hardware, steel siding and roofing, weather vanes, cupolas and other hard-to-find items. They sell and install products.
Phone: 800 836-4271. Fax: 716 457-3160.
Manufactured Buildings, Kits, Custom Design, Cupolas, Roofing, Doors & Windows

Summit Door, Inc.
1233 Enterprise Court, Corona, CA 91720
Summit Door, Inc. manufactures custom sectional garage doors. They offer an unlimited variety of wood species and door designs. Create your own design or choose one of theirs. Call or Fax them your blueprint. Free literature is available. Phone: 888 768-3667. Fax: 909 272-6367. Website: www.summit-door.com.
Sectional Garage and Coach House Doors

Strictly Barns

Sun-North Systems Ltd.
29 Railway Street, Seaforth, Ont. N0K 1W0 Canada
Natural ventilation systems for all agricultural building types and for all climates and geographical areas. Dual and single ventilation products are designed for specific building needs. Free product literature. Phone: 519 527-2470. Fax: 519 527-2560. E-mail: sunnorth@sunnorth.com. Website: www.sunnorth.com.
Barn and Agricultural Building Ventilators

Tech Art
225 Valley River Avenue, Suite G, Murphy, NC 28906
Tech Art offers building plans for small cabins and country stores. Sizes range from 144 sq. ft. to 576 sq. ft. All of the designs can be seen on their website. They also provide design and drafting services for custom designs and a new internet service, *Barns, Barns, Barns*, that presents blueprints by various architects and designers. Phone or Fax: 828 837-4580. E-mail: noff@grove.net. Website: www.grove.net/~noff/.
Barn Plans, Cabin Building Plans, Custom Design

Summit Door, Inc.

Texas Timber Frames
7214 Echert, San Antonio, TX 78238
Texas Timber Frames designs, engineers, joins and raises traditional old-world handcrafted timber frame structures, timber trusses, timber roof systems and architectural timber work. Free literature. Phone: 210 647-4662. Fax: 210 647-4667. E-mail: info@texastimberframes.com. Website: www.texastimberframes.com.
New Timber Frames, Custom Design

Timber & Stone Restorations
5431 East U.S. Highway 290, Fredericksburg, TX 78624
Timber & Stone salvages, restores and resells vintage log structures and timber frames - from small smoke houses to 3,000+sf log barns. They also provide antique material and accessories. Free literature. Video: $10.00. Phone: 830 997-2280. Fax: 830 997-1195. E-mail: magnum@ktc.com. Website: timberandstone.com.
Vintage Log Cabins, Log Barns, and Timber Frames, Antique Building Materials, Restoration Services

Timber Creek Post and Beam Inc.
P.O. Box 309, Cuttingsville, VT 05738
Provides timber frame homes and barns, hand-crafted from eastern white pine using traditional mortise and tenon joinery. Custom design, quality and flexibility are part of Timber Creek's tradition. Free literature. Phone: 802 775-6591. Fax: 802 775-6591. E-mail: timber@sover.net.
New Timber Frames, Custom Design

Timber Creek Post and Beam Inc.

Timber Frames by R.A. Krouse
46 Titcomb Lane, Arundel, ME 04046
Complete traditionally joined barns, homes and other structures, delivered and raised throughout New England. Frames are cut from white pine, selected from Maine's western mountains. Resawn Douglass fir and white oak frames are also available. Call for a house and barn tour. Free literature. Phone or Fax: 207 967-2747. E-mail: rakrouse@cybertours.com. Website: www.mainetimberframes.com.
New Timber Frames, Custom Design, Timber Frame Garden Structures

Timberpeg
PO Box 5474, West Lebanon, NH 03784
Hand-crafted, custom designed timber frame homes and other structures are offered in Douglas fir or pine. Pre-cut packages are fabricated and shipped from east or west coast facilities. Standard designs are also available. Catalog: $15.00. Free literature. Phone: 603 542-7762. Fax: 603 542-8925. E-mail: info@timberpeg.com. Website: www.timberpeg.com.
New Timber Frames, Custom Design

Eli Townsend & Son
132 Hemlock Drive, Deep River, CT 06417
Traditional New England designs of cottages, garages, backyard barns and studios. Reasonably priced plans were prepared by a professional engineer. Free literature. Phone: 860 526-3896. E-mail: john@townsendbooks.com. Website: http://albino.com/Townsend.
Building Plans (See the designs on pages 30 & 53)

United Overhead Door Corp.
21 Saw Mill River Road, Yonkers, NY 10701
Designers and manufacturers of custom wood overhead garage doors and carriage house doors. Free product literature. Phone: 914 964-0038. Fax: 914 964-0964. E-mail: louise@uniteddoor.com. Website: www.uniteddoor.com.
Carriage House Doors, Overhead Garage Doors

United Overhead Door Corp.

Vafac, Inc.
212 Freedom Court, Fredericksburg, VA 22408
Vafac, Inc. sells classic horse equipment. Products include stalls, grills, doors, stall mats, rubber pavers, arena kick board systems, feeders, and automatic watering bowls. Free literature. Phone: 540 898-5425. Fax: 540 898-8442. E-mail: sales@horsestallsusa.com.
Stable Equipment

Vermont Stresskin Panels
184 John Putnam Memorial, Cambridge, VT 05444
Provides stresskin panel enclosure systems for timber frame structures. They provide fast, tight and economical shells for energy efficient buildings. Free literature. Phone: 802 644-8885. Fax: 802 644-8797. E-mail: info@stresskin.com. Website: www.stresskin.com.
Stresskin Panels for Timber Frame Homes

Vermont Timber Frames
7 Pearl Street, Cambridge, NY 12816
Provides traditional timber frame structures for homes, barns, stables and outbuildings. Free literature. Phone: 518 677-8860. Fax: 518 677-3626. E-mail: tomharrison@vtf.com. Website: www.vtf.com.
New Timber Frames, Custom Design

Vintage Barns, Woods & Restorations, Inc.
333 Mossy Brooks Road, High Falls, NY 12440
Specializes in the restoration and reproduction of colonial structures, timber frame barns, outbuildings, log homes and stone structures. Their primary area of service is the middle Atlantic states. They dismantle, re-erect and convert antique frames, and provide resawn siding and flooring. Free literature. Phone or Fax: 914 256-9564. E-mail: barns@hvi.net. Website: www.vintagewoods.com.
Vintage Timber Frames and Log Homes, Restoration Services, Custom Design

Vintage Barns, Woods & Restorations, Inc.

Walters Buildings
PO Box 388, Allenton, WI 53002
Serves the Midwest with high quality pre-engineered agricultural buildings, garages and horse barns. Free product literature. Phone: 800 558-7800. Fax: 414 629-5233. Website: www.waltersbuildings.com.
Garages, Horse Barns, Agricultural Buildings

The Weather Hill Company
PO Box 113, Charlotte, VT 05445
Specialists in classic, traditional home design, restoration, reproduction woodwork and consultation. They work in New England and nationwide. The Westher Hill Company maintains an inventory of historic buildings which may be moved to your property. Free literature. Phone: 802 425-2095. Fax: 802 425-6402.
Restoration Services, Custom Design, Woodwork

Weather or Knot Antiques
8504 West 1350 S, P.O. Box 321, Wanatah, IN 46390
Provides antique and modern lightning rod glass ornaments, antique rods and weather vanes. They offer restoration material for lightning protection memorabilia. Catalog: $10.00. Phone: 219 733-2530.
Lightning Rods, Decorative Glass Insulators, Weather Vanes

West Coast Weather Vanes
377 Westdale Drive, Santa Cruz, CA 95060-9446
West Coast Weather Vanes creates handcrafted, limited-edition copper and brass weather vanes for residential, commercial and public facilities and gardens. Over 300 standard designs are available. Custom and personalized vanes can be commissioned. Free catalog. Phone: 800 762-8736. Fax: 831 425-5514. E-Mail: info@westcoastweathervanes.com. Website: www.westcoastweathervanes.com.
Weather Vanes

West Coast Weather Vanes

Wind & Weather
PO Box 2320, Mendocino, CA 95460
Wind & Weather's catalog offers weather vanes, cupolas and finials for all types of country homes, barns and outbuildings. Products include weather instruments, yard bells and chimes, wind sculptures, sundials and garden ornaments. Free catalog. Phone: 707 964-1284. Fax: 707 964-1278.
Cupolas, Weather Vanes, Garden Accessories

Windy Hill Forge

Windy Hill Forge

3824 Schroeder Ave., Perry Hall, MD 21128
Custom barn door strap hinges, large gate hinges, door hinges, bolts, hasps, cast iron wall washers, restoration work on antique iron hardware. Free product literature. Phone 410 256-5890. E-Mail: windyhillforge@juno.com.
New and Antique Forged Hardware

Woodcraft
P.O. Box 1686, Parkersburg, WV 26102
Woodcraft offers the highest quality woodworking tools including socket slicks, corner chisels, heavy duty framing chisels and adzes. Free catalog. Phone: 800 225-1153. Fax: 304 428-8271. Website: www.woodcraft.com
Woodworking Tools and Books

Woodford Bros., Inc.
PO Box 108, Apulia Station, NY 13020
Structural repair of all structures, specializing in post & beam barn repair. Free product literature. Phone: 800 OLD-BARN. Fax: 315 696-5931. E-mail: mwoodf5685@aol.com.
Restoration Services

Woodhouse - The Timber Frame Company
Box 219, Route 549, Mansfield, PA 16933
and PO Box 1778, Breckenridge, CO 80424
Woodhouse is a custom manufacturer of timber frame house and barn kits. They use premium materials and craftsmanship. Their services range from architectural design through the construction of their packages. Catalog: $17.00. Phone: 800 227-4311. Fax: 570 549-6233. Website: www.woodhouse-pb.com.
New Timber Frame Home and Barn Kits, Custom Design

Wood's Metal Studios
6945 Fishburg Road, Huber Heights, OH 45424
Offers custom forging of traditional and contemporary ironwork, including gates, railings, stair rails, hardware, lighting, etc. Wood's Metal Studios can match your antique hardware. Phone: 937 233-6751.
Reproduction Hardware, Gates & Railings

Vermont Timber Frames

Woodstar Products, Inc.
PO Box 444, Delavan, WI 53115
Woodstar Products, Inc. provides one-stop shopping for all of your horse stall component needs. They offer doors, grills, front panels, hardware, swing out feeders, water buckets, stall mats and other products. Free catalog. Phone: 800 648-3415. Fax: 414 728-1813. E-mail: woodstsr@idcnet.com. Website: www.wdstwr.com.
Stable Equipment

George Yonnone, Restorations
PO Box 278, West Stockbridge, MA 01266
Specialist in timber frame structural repair, jacking, squaring and sill replacement. Yonnone can provide complete interior and exterior restoration, including dismantling and relocating antique structures and log cabins. Free literature. Phone: 413 232-7060.
Vintage Timber Frames and Log Cabins, Restoration Services

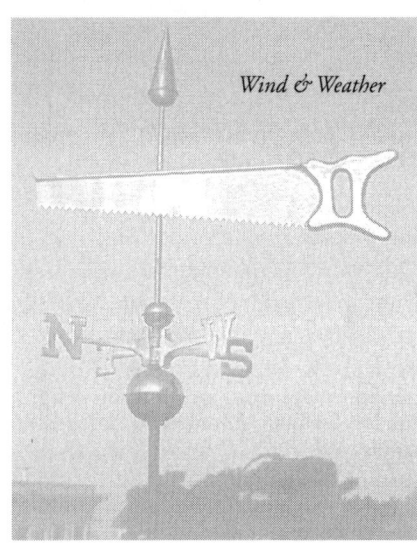

References & Resources

One of the many designs on the new *Barns, Barns, Barns* website.

Internet Resources

The Internet is quickly becoming the best place to research a building project. These websites and e-zines offer reliable information.

BackHome Magazine has a variety of useful information on their website - great books on country building, homesteading and living with nature, articles from the latest issues of the magazine and an on-line catalog of products and tools. You'll also find an index of back issues, with dozens of articles on traditional and unusual country building techniques. The website is www.backhomemagazine.com.

B4UBUILD.COM is an on-line source of information about residential construction and design, the custom home building process, pictures of houses, pet peeves, book reviews, software and a directory of homebuilding resources. You'll find the website at http://www.b4ubuild.com.

Barns, Barns, Barns is a new site operated by Tech Art as a companion to their great *Cabin Plans* website. It offers blueprints of barns, sheds, stables, garages and outbuildings by various architects and designers. Blueprints available on this site are selected by editor Donald Berg for the quality of information that they provide to builders. You can order them directly from the designers, or e-mail them with questions. The site is continuously updated with new plans. See it at www.grove.net/~noff/barns.html.

The Barn Journal On-line is a website dedicated to the appreciation and preservation of traditional farm architecture. Editor Charles Leik reviews new publications and posts news about events and resources. The free classified ads are a terrific way to find restoration specialists, old barn frames and authentic hardware and fittings. Check it out at http://museum.cl.msu.edu/barn.

Barn Again! is a website run by *Successful Farming Magazine* and The National Trust for Historic Preservation. It provides information to help owners of historic barns rehabilitate them and put them back to productive use. The website address is htpp://www.agriculture.com/contents/ba!/ba!.html.

The Cottage Home Network, at www.cottagehome.net, is a valuable resource for anyone planning to build in the country. Jim Tolpin, author of the bestselling book, *The New Cottage Home*, presents home designs, book reviews and links to manufacturers of hard-to-find architectural and restoration products.

D-I-Y Sheds and Barns is the perfect site for do-it-yourselfers. Articles, book reviews and dozens of links to kit manufacturers and blueprint designers make this the best place to start any backyard project. Go to www.geekbooks.com/walden/d-i-ysheds-n-barns.html.

Farmer's Market Online is a great source for farm supplies, books, green house kits, seeds, recipes, herbs, sauces and spices. Log on at wwwfarmersmarketonline.com.

The HayNet is an essential site for anyone who builds for horses. In its "Barn and Farm Equipment" section, you'll find dozens of links to barn designers and stable equipment manufacturers. Sections on horse care books, stable management, grooming equipment and on-line magazines are the most extensive lists of links that you'll find on the Internet. Go to www.haynet.net.

ImproveNet will help you find qualified architects, designers and contractors in your neighborhood and then guide you through the construction process. Their claim of "1,000s" of construction products in their easy-to-search directory is no exaggeration. Check it out yourself at www.improvenet,com.

The Old-House Journal Online, at www.oldhousejournal.com, has articles from the magazine, historic house plans and directory of restoration professionals and products.

Post-Frame Construction, the site run by the National Frame Builders Association, posts a state by state directory of their members, who are professional "pole-barn" and home builders, designers and manufacturers of country building materials. Visit the website: www.postframe.org, to find the pros in your area.

The Timber Framers Guild website is your best introduction to the modern practioners of the ancient craft. Build a timber frame barn and it will last 300 years. The site has informative articles and news of its members' projects, activities and events. Links and e-mail addresses will help you find professionals in your area to help you design and build your barn. The site is at www.tfguild.org.

Timber Framing Magazine, has information on timber framing and alternative building. The website is www.timberframingmagazine.com.

Traditional Building is the website run by the publishers of *Traditional Building* and *Period Homes* magazines. Its search engine and directory is the best on-line source of restoration products and services. Find it at www.traditional-building.com.

Building Codes

Your region is probably covered by one of the four major building codes. Call or write to get information or to order code books.

Building Officals & Code Administrators International (BOCA)
4051 West Flossmore Road, Country Club Hills, IL 60477
708 799-2300

International Code Council (ICC)
5203 Leesburg Pike, Falls Church, VA 22041
703 931-4533

International Conference of Building Officials (ICBO)
5360 South Workman Mill Road, Whittier, CA 90601
562 699-0541

Southern Building Code Congress International (SBCCI)
900 Montclair Road, Birmingham, AL 35213
205 591-1853

Building Book Catalogs

These organizations offer free catalogs full of hard-to-find books on building and woodworking specialties. Call or write them for your copy or visit their websites.

Builder's Booksource
1817 4th Street, Berkeley, CA 94710
Free Catalog. Phone: 510 845-6874.

Linden Publishing
336 West Bedford, #107, Fresno, CA 93711
Free *Woodworkers Library* catalog has over 300 books and videos.
Phone: 800 345-4447. Fax: 559 431-2327. Website: www.lindenpub.com.

Home Builders Bookstore of the National Association of Home Builders
1201 15th Street, NW, Washington, DC 20005-2800
Free catalog offers guide books to help you work well with your home builder or remodeler. Resources on caring for your home and a wide variety of professionsl reference and "how-to" books for home builders.
Phone: 800 368-5242. Website: builderbooks.com.

Summer Beam Books
2299 Route 488, Clifton Springs, NY 14432
Free catalog of books on timber framing, woodworking, barns, house design and construction and related crafts.
Phone: 877 279-1987, toll free. Fax: 716 289-3221. E-mail: char@fltg.net. Website:

Periodicals

Historic Homes & Properties is a monthly tabloid supplement to *Antiques & The Arts Weekly*. It features historic homes and vintage properties. It has fascinating articles, great book reviews and sources for salvaged architectural materials, barn frames and restoration specialists. Call for information on subscriptions: 203 426-3141.

This Old House Magazine has informative articles on restoring and maintaining old buildings. It takes a practical approach to an impractical endeavour. If you have an old carriage barn, you'll need a subscription. Call 800 898-7234.

The Old-House Journal and its annual *Restoration Directory* are essential sources for restoration products and advice. The magazine publishes an *Historic House Plans* issue twice a year, and that usually features some traditional designs of garages and outbuildings. For subscriptions or copies of the special issues, call 800 931-2931.

Barn & Backyard Building Guidebooks

Building Fences & Gates: How to Design and Build Them from the Ground Up, by Richard Freudenberg (Lark Books, 1997). A great source for design ideas and advice on planning, plotting and building your fence.

Building Fences of Wood, Stone, Metal, & Plants, by John Vivian (Williamson Publishing, 1992). This is a general primer on building all types of fences and on growing hedges. The section on stone and masonry is particularly thorough. Great illustrations by Liz Buell, straightforward text and detailed photographs of works in progress make this a good resource for both novice and experienced builders.

Building a Multi-Use Barn, by John D. Wagner (Williamson Publishing, 1994). Builder John Wagner shows the versatility that's possible with one good barn. Using a 24' by 30' plan and simple framing, Wagner alters the interior layout to create a tractor garage and garden shed, a studio, a workshop and office, and a stable. His ideas should be considered by anyone looking for practical uses for old barns and heritage frames. Besides being a design guide, this book covers all the basics of light frame construction with easy-to-read text, photos and great illustrations.

Building Small Barns, Sheds & Shelters, by Monte Burch (Story Books, 1983). From permits to the finish coat of paint, this book will guide you through your building process. Burch describes the advantages of different framing methods, roof styles and materials and backs his text with useful reference tables and concise construction details. The book presents plans for five small barns, two two-stall stables, a root and storm cellar, a carport, a tool shed, a woodshed, a smokehouse and shelters for hens, pigs and rabbits.

Complete Plans for Building Horse Barns Big and Small, by Nancy W. Ambrosiano and Mary F. Harcourt (Breakthrough Publications, 1997). If you're planning to design or build a stable, you'll find yourself using this book again and again. It's a countrywide survey of creative designs for equestrian buildings. The buildings are presented with plans, photos and concise descriptions.

Horsekeeping on a Small Acreage: Facilities Design and Management, by Cherry Hill (Storey Books, 1990). This common-sense guide, by a horse care expert, has information on planning your property, building design, fences, paddocks, fire safety, pasture and hay-lot management and much more. The book is packed with photos and with illustrations, plans and details of stable designs.

How to Build Small Barns & Outbuildings, by Monte Burch (Storey Books, 1992). Burch combines great building advice with plans for 20 small buildings. You'll find designs for three small all-purpose barns, an eight-stall horse barn, various animal shelters, two garages and four garden sheds.

Practical Pole Building Construction, by Leigh Seddon (Williamson Publishing, 1985). A complete builders' guide with reference tables, over 100 clear illustrations, photos and building plans for a lean-to animal shelter, a two-stall stable, a combination two-car garage and woodshed, and more.

Pole Building Projects, by Monte Burch (Storey Books, 1993). This book presents the basics of pole building and design and includes plans you can build for barns, sheds, garden structures and garages. Useful tables, charts, photos and illustrations form a step by step guide.

Roofs and Rails: How to Plan and Build Your Ideal Horse Facility, by Gavin Ehringer (Western Horseman, 1995). Ehringer covers all aspects of horse barn design and construction, from planning your acreage to hanging a halter. This book is filled with photos and easy-to-understand plans and details.

Rustic Retreats: A Build-It-Yourself Guild, by David & Jeanie Stiles (Storey Books, 1998). Straightforward instructions and beautiful, informative drawings will help you build dozens of great back-country shelters. Designs include sheds, arbors, lean-tos, huts, cabins, tree houses and even a design for a floating cabin.

Sheds: The Do-it-Yourself Guide for Backyard Builders, by David Stiles (Firefly Books, 1998). This book has the information you need to design and build your own ideal backyard shed. In fact, it serves as a great primer for any construction project. It covers planning, designing, permits, materials and construction methods. Stiles will guide you through the process, step-by-step, from the paper plan to hanging up your tools - in the shed you built! There are hundreds of great illustrations and projects you can try: a Victorian garden shed, cupolas, a Japanese boat shed, trash and recycling sheds, woodsheds, a pool pavilion and many more.

Step-by-Step Outdoor Stonework, by Mike Lawrence (Storey Books, 1998). Twenty different projects are presented in color photos and detailed drawings. Lawrence concentrates on patios, paving, steps, garden walls, stone furniture and decorative ponds. These are all projects that homeowners can handle themselves with this concise guide.

Stonework Techniques and Projects, by Charles McRaven (Storey Books, 1997). A guide to the basics of stonework that concentrates on the most common projects: retaining walls, stone fences, foundations and steps, and then adds a bit more for the adventuresome: a fireplace, an arched bridge and a moon gate. A good book for the do-it-yourselfer or to learn what to look for in hiring a professional stone mason.

Illustration from
How to Build in the Country

Books on Yesterday's Designs

American Barns, by Stanley Schuler (Schiffer Publishing, 1984). Take a tour of 240 old and new barns throughout the United States with this book's clear photographs and concise descriptions. You'll have a good introduction to our regional styles and to the amazing variety of different barn types. If you're planning to design or build a barn, you can't help but be inspired by the many photos of building details.

American Country Building Design: Rediscovered Plans for 19th Century Farmhouses, Cottages, Landscapes, Barns, Carriage Houses & Outbuildings, by Donald J. Berg, AIA (Sterling Publishing, 1997). Some of yesterday's best designs are shown in original engravings, plans and in the words of their designers. Historic woodwork details and site planning techniques can help you design and build in the American country tradition

Barns, by Charles Leik (Friedman/Fairfax, 1999). Incredible color photographs of barns from all regions and an eloquent introduction by Leik make this book of old barns an inspiration for anyone contemplating a building or restoration project.

Barns, Sheds & Outbuilding: Placement Design and Construction, edited by Byron D. Halstead (Alan C. Hood & Company, 1994). This is a direct reprint of an 1881 classic, with a new forward by Castle Freeman. Halstead selected some of the most popular published designs from over two decades of the farm journal, *The American Agriculturist*. Designs include barns, stables, carriage houses, animal shelters, corn cribs, ice houses, spring houses, dog houses and bird houses.

How to Build in the Country: Good Advice from the Past on How to Choose a Site, Plan, Design, Build, Decorate & Landscape Your Country Home, edited by Donald J. Berg, AIA (Donald J. Berg, AIA, 1999). The long-lost tricks of the trade that created yesterday's picture-perfect country homes are available again. They can help you build your dream home. This book has 178 hints, 158 old-time engravings and 150 years worth of wisdom from the best of yesterday's architectural books, home magazines and farm journals.

The Farm: an American Living Portrait, by Joan and David Hagan (Schiffer Publishing, 1990). Hundreds of color photographs document the American family farm and its passing way of life. If you're planning a barn building project, this book's crisp shots and close-up details of beautiful barns, outbuildings and cupolas are sure to inspire you.

Other Source Books by Donald J. Berg, AIA

BARNS AND BACKBUILDINGS: Designs for Barns, Carriage Houses, Stables, Garages & Sheds, with Sources for Building Plans, Books, Timber Frames, Kits, Hardware, Cupolas & Weather Vanes.

If you're planning to build in your backyard, you need this book. It has 94 variations of designs for barns, stables, carriage houses, craft barns, garages, workshops, studios, sheds, play-houses and garden structures. You'll find plans, kits, timber frames and prefabricated structures. You'll also find some common-sense advice on planning and building on your property.

Rebuild a historic, hundred year old timber frame barn and preserve part of America's rural heritage. Have a new frame built just for you. Find the architect or designer who can create a custom building for your special needs. Find the mail-order plan that's just right and then add a pre-built cupola, hand-forged hardware and a weather vane custom made just for you. Now you have the sources to turn your property into the perfect country place.

"Charming Outbuildings" - *The Baltimore Sun.* "A must book if you are planning the construction of a small barn" - Charles Leik, Editor, *The Barn Journal On-Line.*

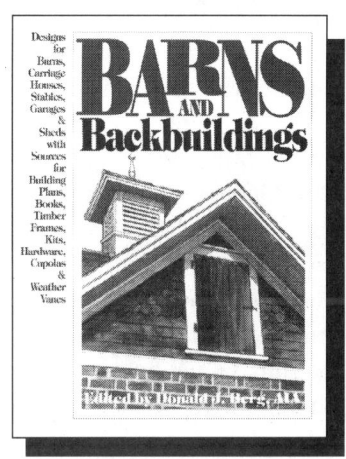

Edited by Donald J. Berg
96p, 8 1/2 x 11, paperback, $9.95
ISBN 096630750X

THE BACKROAD HOME: Simple Country Designs of Cottages, Cabins, Barns, Stables, Garages and Garden Sheds, with Sources for Blueprints, Kits, Building Accessories, Catalogs and Guide Books

Build a simpler life. This sourcebook presents 86 small, simple and easy-to maintain cottages, cabins and outbuildings by seventeen of America's most talented country architects and designers. It has their addresses, phone numbers and websites plus a directory of 217 more catalogs and websites for blueprints, kits, guidebooks and hard-to-find country building products.

"If you want to get started right away on your cottage project with some really attractive, off-the-shelf designs, this is the first book you should buy." - Jim Tolpin, *The Cottage Home Network.* "If you have hopes and plans to build that tiny house 'by the side of the road,' you should add this informative book to your library." - *Historic Homes & Properties*

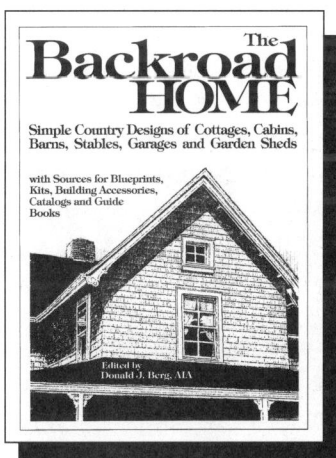

Edited by Donald J. Berg
96p, 8 1/2 x 11, paperback, $9.95
ISBN 0966307526

Barns & Backbuildings ($9.95), *The Backroad Home* ($9.95), *How to Build in the Country* ($12.95), and additional copies of *Carriage Barns* ($9.95) are available at libraries and bookshops, from Internet bookstores, or by mail from Donald J. Berg, AIA, PO Box 698, Rockville Centre, NY 11571. Please send payment, and $3 postage (the $3 covers any quantity of books ordered). The books come with a money-back guarantee. To order by phone, call **800 887-2833** and leave a recording with your name, address and credit card information.

Country Building Books 95

Index

Antique Timber Frames - see Vintage Timber Frames
Apartments - see Live-Ins
Barn Doors - see Carriage House Doors
Barns, 24,26,27,32-38,41-44,50-52,63,89,90,95
Blueprints - see Building Plans
Building Codes, 91
Building Guide Books, 91-94
Building Kits, 31,40,52,70,72,74-76,79-83,86,87,95
Building Materials, 67-88
Building Plans, 23-38, 40-44,46-56,58-66,68-73,75-85,89,90,95
Carriage House Doors, 74,76,84,85
Carriage House Lamps, 72
Carriages, 18,19,83
Cupolas, 66,67,69,71-75,78,79,81,84,87
Custom Design, 67-69,74,75,78-87
Finials, 68,72-75
Garage Apartments - see Live-Ins
Garages, 24-38,41,46-65
Garden Buildings, 36,41,70,74,75,77,90,95
Gazebos - see Garden Buildings
Greenhouses - see Garden Buildings
Guest Houses - see Live-Ins
Hardware, 67-71,73-78,82,87
Heaters, 78
Historic Plans - see Yesterday's Designs
Home Offices, 30,45-56,58-65
Home Plans - see Building Plans
House Kits - see Building Kits
Internet Resources, 89,90

Lightning Rods, 77,81,86
Live-Ins, 45-56
Magazines, 92
Mail-Order Plans, see Blueprints
On-Line Resources, 89,90
Outhouses, 70
Pole Barns, 24,25,27-29,35,68,90
Property Plan, 22
Resources, 89-95
Restoration, 7,67-74,77-82,84-90
RV Garage, 34
Scale of Plans, 7
Sheds, see Garden Buildings
Site Plan, 22
Stables, 36-44,72,86
Stable Equipment, 76,78,80,82,83,85,88,90
Stairs, 69,76,83,84
Studios, 24,25-30,31,33,46-56,58-65
Timber Frames, 69,71-75,79-81,83-86,90
Vacation Homes - see Live-Ins
Ventilators, 72,78,81,84
Vintage Timber Frames, 67-75,79-81,85,86,89
Weather Vanes, 67,71-75,77-79,86,87
Woodwork, 71,73-75,79,80,82,86
Woodworking Tools, 87
Working Drawings - see Building Plans
Workshops, 24,25,27-29,31,32,35-37,44,51,56,58-61
Yard Bells, 71,73
Yesterday's Designs, 8-23,39,94